Visual Quality Assessment for Natural and Medical Image

Yong Ding

Visual Quality Assessment for Natural and Medical Image

Yong Ding
Zhejiang University
Hangzhou
China

ISBN 978-3-662-58583-2   ISBN 978-3-662-56497-4   (eBook)
https://doi.org/10.1007/978-3-662-56497-4

Jointly published with Zhejiang University Press, Hangzhou

The print edition is not for sale in China Mainland. Customers from China Mainland please order the print book from: Zhejiang University Press.

This Springer imprint is published by Springer Nature
The registered company is Springer-Verlag GmbH, DE
The registered company address is: Heidelberger Platz 3, 14197 Berlin, Germany

Preface

With the rapid development of digital image and video acquisition, transmission, and display techniques in the last few decades, the demands of high-quality images and videos are growing amazingly fast in both people's everyday lives and specific application scenarios such as the fields of academy and engineering. Image quality assessment (IQA) is an essential technique in the design of modern large-scale image and video processing systems. Aiming to evaluate the quality of images objectively, effectively, and efficiently, IQA serves as performance evaluator and monitor of these systems and helps them with the pursuit of functional improvement. As a result, study upon IQA is attracting more and more attention in recent years and a significant progress has been witnessed.

As far as the author is concerned, this book is the first to thoroughly discuss the related topics about IQA, including the basic principles of subjective and objective experiments, biological evidence for image quality perception, recent researching developments. Especially, to cover the recent trends of imaging techniques and to explain the application-specific utilization of the field, extra efforts have been spent in this book in introducing IQA for stereoscopic (3D) images and medical images, rather than a mere focus on planar (2D) natural images.

The author would give particular thanks to Yang Zhao, Xinyu Zhao, Xiaobao Shang, Shaoze Wang, and Ruizhe Deng who have made a significant contribution to the publication of this book. Without their strong support, this book could not have been finished.

The author would also like to express deep appreciation to my Ph.D. and master students who contribute to the research work presented in this book. The author would like to thank them for their great efforts in researching on IQA for the last several years. For example, Yuan Zhang is the first person in our group to branch out tentatively into this interesting topic. And Hang Zhang, Shaoze Wang, Hang Dai, Xiang Wang, Yifan Zhu, and a number of graduate students devote their energy and enthusiasm to it further.

Besides, the author has received generous assistance and support from many of our colleagues including valuable information and materials used in the book, discussions, feedback, comments on and proofreading of various parts of the book, recommendations, and suggestions that shaped the book as it is. Special thanks are due to Xiufang Wu and Tianye Niu.

Due to our limited knowledge and energy, there inevitably exist some ambiguous interpretations and even mistakes in this book, which we welcome the readers and colleagues to point out.

Hangzhou, China   Yong Ding
September 2017

Contents

Chapter 1
Introduction

With the increasing demand in applications as broad as entertainment, communications, security, monitoring, and medical imaging, visual information (image or video) plays a more and more important role in human's daily life. However, the visual quality of image may suffer from potentially substantial loss during procedure of image acquisition, processing, compression, transmission, and reproduction (Wang and Bovik 2009; Karam et al. 2009; Chang et al. 2015; Saha and Wu 2016). Therefore, how to evaluate the image quality accurately has become a hot topic in recent years. Image quality assessment (IQA) is essential not only on its own for testing, optimizing, and inspecting related algorithms and image processing systems (Sheikh et al. 2005; Zhang and Chandler 2013; Wu et al. 2016a, b), but also for shaping and decision making for virtually all multimedia signal processing and transmission algorithms (Deng et al. 2015).

Since human visual system (HVS) is the ultimate receiver of images, ideally, IQA can be conducted using subjective tests in which human subjects are asked to rate the perceived visual quality according to a provided quality scale and specified criteria. Subjective IQA completed by human observers always reflects the perceptual quality of images loyally, yet it is time consuming, cumbersome, and unstable, resulting in its impracticability to be applied in real-time systems (Larson et al. 2010; Gao et al. 2013; Oszust 2016). It triggers the urgent need to develop reliable objective IQA metrics which can automatically measure the perceptual image quality that is consistent with subjective human evaluation (Mittal et al. 2012; Zhang et al. 2014; Chang et al. 2015; Wu et al. 2016a, b).

With the rapid development of the sensing and imaging devices, newly emerged visual signals are presented to human viewers, such as 2D image, stereoscopic/3D image, graphics, medical image. Thus, the image quality assessment is roughly extended from 2D natural image quality evaluation to several categories. Meanwhile, recent psychophysical and neurological findings enable us to more clearly understand the human visual system. All these progresses make the research field of image quality assessment experience significant growth during the last decade (Deng et al. 2015).

The objective of this book is to provide a comprehensive review of recent advances of image quality assessment and shape the future research directions. However, not all aspects of such a large field of study can be completely covered in a single book; therefore, we have to make some choices. Basically, we concentrate on 2D natural image quality assessment, stereoscopic/3D image quality assessment, and medical image quality assessment. Each chapter begins with an introductory section and includes an overview of developments in the particular area of research. And the discussed contents in each chapter or section are expected to help not only inspire newly research trends and directions for IQA but also benefit the development of multimedia products, applications, and services. Furthermore, many of the citations at the end of each chapter are from recent work published in the literature.

Chapter 2 gives a brief overview of subjective ratings and image quality databases. In order to evaluate the performance of an objective IQA method, ground-truth image quality databases are necessary. To build up an IQA database, a set of images are shown to a group of human observers who are asked to rate the quality on a particular scale. The mean rating for an image is referred to as the mean opinion score (MOS) and is representative of the perceptual quality. Then, the score predicted by the IQA method is correlated with MOS; a higher correlation is indicative of better performance. In recent years, numbers of image quality databases annotated with subjective quality ratings have been published and are publicly available (Winkler 2012).

Chapter 3 is a foundational introduction of human visual system (HVS), though the knowledge of it is far from complete. Since HVS is the ultimate receivers and processors of image, understanding and modeling the perceptual mechanism is significant for IQA development. Firstly, the basic structures of HVS are introduced. Then, the typical properties of HVS applied in current IQA implementations are discussed.

Chapter 4 provides a rough introduction about the general framework of modern image quality assessment, where the typical schemes of three categories, full-reference (FR), reduced-reference (RR), and no-reference (NR) are given, respectively. Furthermore, the stages of the IQA framework including quality-aware feature extraction, feature quantification, and quality mapping strategy are discussed.

Chapter 5 focuses on IQA methods based on human visual system properties. Since images are ultimately viewed by human beings, modeling the way that human beings perceive an image is a meaningful solution for image quality assessment. Based on the understanding on the complex and rigorous HVS, numbers of IQA methods are proposed either draw inspirations from the HVS hierarchies or the responses of HVS. In addition, the IQA methods based on visual attention (saliency) are discussed.

Chapter 6 presents a survey and discussion on the image quality assessment based on natural image statistics. Since it is very difficult to model the complex and rigorous HVS well relying on the limited understanding upon it, most of state-of-the-art IQA methods attempt to extract the statistical properties (features) of an image that are closely related to the inherent quality. In this chapter, methods based on structural similarity, multifractal analysis, textural features extraction, and independent component analysis are discussed.

Chapter 7 addresses the stereoscopic image quality assessment which is different from the traditional 2D IQA. In this chapter, firstly, the binocular vision is reviewed briefly. Then, subjective stereoscopic IQA and existing databases are introduced. And finally, detailed discussions about current objective stereoscopic IQA methods are provided.

Chapter 8 reviews recent progresses of quality assessment for medical images briefly and then concentrates on presenting a quality assessment method for portable fundus camera photographs, putting forward a generalized relative quality assessment scheme, further giving an adaptive paralleled sinogram noise reduction method for low-dose X-ray CT based on the proposed relative quality assessment scheme, and finally studying on the relationship between the image quality and imaging dose in low-dose CBCT based on dose-quality maps.

Chapter 9 discusses challenge issues and new trends of image quality assessment in the future.

This book is suitable for researchers, clinicians, and engineers as well as students working in related disciplines including imaging, displaying, image processing, storage, and transmission. It is believed that the review and presentation of the latest achievements, new trends, and challenges in image quality assessment will be helpful to the researchers and readers of this book.

References

Chang, H., Zhang, Q., Wu, Q., & Gan, Y. (2015). Perceptual image quality assessment by independent feature detector. *Neurocomputing, 151*(3), 1142–1152.

Deng, C., Ma, L., Lin, W., & Ngan, K. N. (2015). *Visual signal quality assessment*. Switzerland: Springer International Publishing.

Gao, X., Gao, F., Tao, D., & Li, X. (2013). Universal blind image quality assessment metrics via natural scene statistics and multiple kernel learning. *IEEE Transactions on Neural Networks and Learning Systems, 24*(12), 2013–2026.

Karam, L. J., Ebrahimi, T., Hemami, S. S., & Pappas, T. N. (2009). Introduction to the issue on visual media quality assessment. *IEEE Journal of Selected Topics in Signal Processing, 3*(2), 189–190.

Larson, E. C., Chandler, D. M., & Damon, M. (2010). Most apparent distortion: Full-reference image quality assessment and the role of strategy. *Journal of Electronic Imaging, 19*(1), 1–21.

Mittal, A., Moorthy, A. K., & Bovik, A. C. (2012). No-reference image quality assessment in the spatial domain. *IEEE Transactions on Image Processing, 21*(12), 4695–4708.

Oszust, M. (2016). Full-reference image quality assessment with linear combination of genetically selected quality measures. *PLOS ONE, 11*(6), 0158333.

Saha, A., & Wu, Q. M. J. (2016). Full-reference image quality assessment by combining global and local distortion measures. *Signal Processing, 128,* 186–197.

Sheikh, H. R., Bovik, A. C., & Veciana, G. (2005). An information fidelity criterion for image quality assessment using natural scene statistics. *IEEE Trans Image Processing, 12,* 2117–2128.

Wang, Z., & Bovik, A. C. (2009). Mean squared error: Love it or leave it? A new look at signal fidelity measures. *IEEE Signal Process Mag, 1*, 98–117.

Winkler, S. (2012). Analysis of public image and video databases for quality assessment. *IEEE Journal of Selected Topics in Signal Processing, 6*(6), 616–625.

Wu, J., Lin, W., Shi, G., Li, L., & Fang, Y. (2016a). Orientation selectivity based visual pattern for reduced-reference image quality assessment. *Information Sciences, 351*, 18–29.

Wu, Q., Li, H., Meng, F., Ngan, K. N., Luo, B., & Huang, C., et al. (2016b). Blind image quality assessment based on multichannel feature fusion and label transfer. *IEEE Transactions on Circuits and Systems for Video Technology, 26*(3), 425–440.

Zhang, Y., & Chandler, D. M. (2013). No-reference image quality assessment based on log derivative statistics of natural scenes. *Journal of Electronic Imaging, 22*(4), 1–23.

Zhang, L., Shen, Y., & Li, H. (2014). VSI: A visual saliency-induced index for perceptual image quality assessment. *IEEE Transactions on Image Processing, 23*(10), 4270–4281.

Chapter 2
Subjective Ratings and Image Quality Databases

2.1 Introduction

Visual perception information plays a crucially important role in our daily life. In addition, the technology associated with images has changed drastically over the past one hundred years. With the development of the mobile multimedia technology, people can obtain many pictures at any time with high resolutions using their mobile phone or other equipment. However, image system consists of many processes, such as image acquisition, reproduction, compression, storage, transmission, and restoration, which may introduce the noise into images. Therefore, it is significant to monitor and evaluate the image processing.

Ground-truth information is one of the most crucial and essential components for training, testing, and benchmarking of a new algorithm. In the image quality assessment field, ground-truth means image quality databases which generally include a set of reference and distorted images and average quality ratings for each corrupted image. In recent years, numbers of image quality databases annotated with subjective quality ratings have been published for evaluating and refining objective image quality assessment algorithms (Winkler 2012). More than twenty databases for image quality assessment are publicly available in the public domain at present. In this chapter, we review very briefly the subjective image quality rating method and the main publicly available image quality databases.

The rightful judges of visual quality are human because human beings are the ultimate receivers of the visual information. In the meanwhile, the human evaluation results can be obtained by subjective experiments (Shahid et al. 2014). Such experiments try to straightforwardly collect the image quality ratings from the representative pool of a panel of test subjects on the distorted images in the database. It is essential for us to create a controlled laboratory environment to conduct the subjective experiments. Deliberate planning and several other factors have to be taken into consideration in advance to conduct the subjective experiments, such as source content (i.e., reference images), selection of test material (i.e., samples

© Zhejiang University Press, Hangzhou and Springer-Verlag GmbH Germany 2018
Y. Ding, *Visual Quality Assessment for Natural and Medical Image*,
https://doi.org/10.1007/978-3-662-56497-4_2

processed by different test conditions), viewing conditions, grading scale, and timing of presentation. For example, Recommendation (ITU-R) BT.500 (ITU 2012) offers detailed guidelines for conducting the subjective television image quality assessment experiments. Different types of subjective methods for experiment have been developed on various public image quality databases. These subjective methods usually contain single-stimulus-based methods and double-stimulus-based methods. With respect to single stimulus-based methods, the test subjects are only shown test images with different distortions, in the absence of reference images. Of course, in some conditions, a hidden reference image may be contained in the test images; however, the evaluation is just based on the no-reference quality scores of the test subjects. In comparison, double stimulus-based methods mean that the subjects are shown test images along with their corresponding references. The results of the subjective image quality assessment experiments are the quality scores provided by the test subjects, which are employed to calculate the mean opinion score (MOS), differential mean opinion score (DMOS), and another statistics information. In general, the computed values of MOS and DMOS denote the ground-truth information to develop the objective image quality assessment algorithms.

2.2 Subjective Ratings

For subjective image quality assessment method, human subjects are required to assess the image quality under test conditions. Subjective evaluation method is the most reliable approach to determine the quality of real images. There are numerous various methodologies and rules for designing and defining the subjective ratings. In the next, several typical subjective rating methods will be introduced.

2.2.1 Participants—Subjects

People are usually accustomed to estimate objects from various aspects they encounter. It is proved that judgements made by various people tend to be very close (Shahid et al. 2014). With respect to quality of images, people especially reach an agreement on the visual perception information such as the brightness, contrast sensitivity, and visual masking of the images, which is because the peripheral senses of different people are very similar. Meanwhile, this opinion on visual perception information is uninhibited for people having their own opinions about more cognitively related to the same things. On the one hand, everyone has their own aesthetical quality of same things. On the other hand, many various often personal reasons lead to consideration that some images can better give the essence of things. In order to eliminate the deviation and produce reliable results, numbers of people should be participated in the experiment. The researchers usually accepted that 16–24 participants will be able to offer a reliable statistical result. In general, the participants can be classified into experts and nonexperts according to their experience in the image field. All participants should be screened for visual defects, such as visual acuity, color blindness, or color weakness (Wu and Rao 2006).

(1) Experts

Experts usually represent the researchers with rich experience in the image field, such as image evaluation, image generation, image distribution, image compression, and image designing systems. The experts have a special point of view when they observe the images. Employing the experts can become a quick test way, and the experts know what they are working. However, some experts cannot free themselves from their own mode of thinking. Experiments from the experts can be beneficial for the development of the image processing algorithms; however, it may be bad for generalization to the common people. It is an advantage for using experts that they are proficient in offering the technical performance assessments of images. In addition, the experts perform well in free form viewing.

(2) Nonexperts

Nonexperts usually denote the common people, domestic consumer, or someone who will use the product. Sometimes nonexperts can find the artifacts contained in the images that the expert might miss. When the observation images are structured properly, the nonexperts usually can provide the satisfactory experimental results. When selecting the subjects, we should make great efforts to select a representative sample, because the age, gender, and other factors may affect the experiment results.

2.2.2 Experimental Design

There is a potential possibility for every variable in the test to impact the results of image quality evaluation. It is important to consider variety of situations and multiple types of images when designing the experiment.

For subjective image quality assessment, the source signal is very important because it provides the reference image directly and is the input of the test under examination. The stability of the experimental results will be directly affected if there is a flaw in the reference image. Digitally stored images are the ideal source signal because they are reproducible and can be exchanged between the different laboratories.

When conducting a subjective image quality assessment experiment, some factors should be controlled, which contains carefully considering the number of the images and the distortion types of images. It should be mentioned that an experimenter should complete these materials in ninety minutes. The sessions are generally divided into thirty minutes sections. In addition, the experimental structure will be affected by the time duration. A typical presentation diagram is shown in Fig. 2.1, where the training images show the range and types of the distortions. It should be mentioned that the images which are demonstrated to the subjects in the training images time are illustrating images other than those employed in the test. The goal of the training images is to make the subjects familiar with the experimental environment. Moreover, in order to make the opinion of the subjects more stable, five "dummy presentations" should be introduced at the beginning of the experiment. The experimental data of these five presentations is not taken into account in the experimental results.

2.2.3   Presentation of the Result

The experiments will generate distributions of integer value, e.g., between 1 and 5 or between 0 and 100. There will be some little variations in all distributions because of the different judgements between the subjects and the effect of the conditions. However, the results of the experiments are generally reported in the form of mean opinion scores (MOS) or differential mean opinion scores (DMOS). Therefore, the raw data from the trial should be processed.

An experiment will contain numbers of presentations P. In addition, numbers of conditions C will be included in one presentation. Moreover, numbers of images I will be applied in a test condition. Under some circumstances, each combination of test image and test condition can be repeated many times R.

Mean score MOS_{ickr} is one of the most important embodiments of the experimental results, which can be calculated by the following:

$$MOS_{icr} = \frac{1}{N} \sum_{k=1}^{N} S_{ickr} \qquad (2.1)$$

where S_{ickr} denotes the score of subject k for test condition c, image i, repetition r, and N is the number of subjects. Overall mean scores can be computed by the same way.

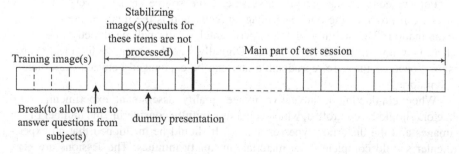

Fig. 2.1 Presentation diagram of test session

Mean opinion score should have an associated interval when representing the experimental results. Ninety-five percentage confidence interval is usually used, which is calculated by:

$$[\text{MOS}_{icr} - \delta_{icr}, \text{MOS}_{icr} + \delta_{icr}] \qquad (2.2)$$

$$\delta_{icr} = 1.96 \frac{S_{icr}}{\sqrt{N}} \qquad (2.3)$$

$$S_{icr} = \sqrt{\sum_{k=1}^{N} \frac{(\text{MOS}_{icr} - S_{icrk})^2}{(N-1)}} \qquad (2.4)$$

A raw data is taken as an outlier if the data is outside the confidence interval. For any session, if a number of values of an observer are outliers, all quality assessments of the observer will be rejected from the obtained raw data. The algorithm of outlier will be run twice.

In order to compute the DMOS, firstly, the raw scores should be converted into quality difference scores, as shown in the following

$$d_{mn} = S_{mref(n)} - S_{mn} \qquad (2.5)$$

where S_{mn} denotes the score of subject m, image n, and $S_{mref(n)}$ is the score of the reference image corresponding to the impaired image n, which is obtained by the subject m. Then, after the reprocessing of outlier removal and experimenter rejection in the same way as above, the difference scores are converted into Z scores,

$$z_{mn} = (d_{mn} - \bar{d}_m)/\sigma_m \qquad (2.6)$$

where $\bar{d}_m$ denotes the mean difference score of all images rated by the subject m and σ_m represents the standard deviation. Then, for the nth image, the Z scores z_n are obtained by calculating the average of z_{mn} across all subjects. By minimizing the error between the DMOS($\bar{z}_n$) and DMOS$_n$ which is obtained from the realignment study, we can learn two values p_1 and p_2. For acquiring the DMOS for the whole database, a linear mapping between Z scores and DMOS is assumed, and the linear mapping is given as follows:

$$\text{DMOS}(\bar{z}) = p_1\bar{z} + p_2 \qquad (2.7)$$

2.2.4 Test Methods

The test methods introduced in this section are internationally recognized and employed by the private enterprise, public sector, and standards organizations to carry out the subjective tests (Choe et al. 1999). It is must careful about each method that guiding the participants discreetly to ensure the subjects can perform the task as required.

(1) Double-Stimulus Impairment Scale Method (DSIS)

Impairment means that a visible degradation is contained in the image. And the task of the subjects is to evaluate the impaired images. A five-grade scale impairment is usually employed to conduct the experiments. The subjects will vote one in the five options. The five options depicted in Fig. 2.2 are imperceptible, perceptible but not annoying, slightly annoying, annoying, and very annoying. The five-grade scale is internationally accepted and recognized for employing in the subjective image quality assessment. The common arrangement of the experimental system is usually shown in Fig. 2.3.

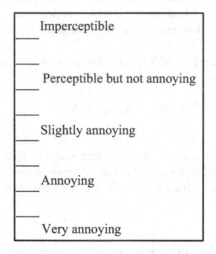

Fig. 2.2 Five-point impairment rating scale

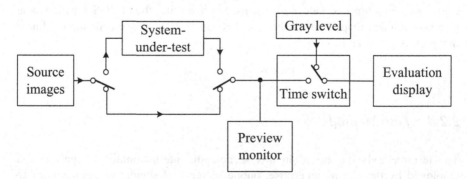

Fig. 2.3 General arrangement for test system of DSIS method

It is a cyclic process for double-stimulus method in which the observers are first observing an unimpaired reference image, and then the same impaired image will be provided. In this case, the observer should vote on the second appeared image with the first image being in mind. As shown in Fig. 2.3, the source image can directly reach to the time switch or indirectly reach the switch via a system under examination. The test images are usually arranged in pairs so that the first image in the pair is directly from the source images, and the second image is the same image through the system under examination. The sessions will last half an hour, and the subjects will observe a series images with random order and impairments covering all the required combinations. Finally, the mean score is calculated.

The reference image implies the image with the highest quality which is usually used to be as the benchmark. The subjects are asked to compare the test image with the reference image and judge the visibility and severity of the impairments introduced into the images based on the five-grade scale mentioned above (Deng et al. 2015). Numbers of presentations will be contained in a test session. The reference image and the test image are presented only once is one variant to conduct the presentation, as shown in Fig. 2.4a, the reference image and the test image are presented twice is other variant to conduct the presentation, as shown in Fig. 2.4b. It is more time consuming for the second variant. In general, studies apply the first variant.

At the start of each session, a detailed guideline is provided to every subject about the detailed information of trial, such as the type of the evaluation, the grading scale, and the timing (reference images, gray, test images, and voting period) (Ponomarenko et al. 2009). The subjects are asked to observe the images for the duration of $T1$ and $T3$ and vote during the $T4$.

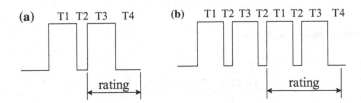

Phases of presentation:
$T1=10s$ Reference image
$T2=3s$ Mid-gray produced by a video level of around 200mV
$T3=10s$ Test condition
$T4=5-11s$ Mid-gray
      Experience suggests that extending the periods T1 and T3 beyond 10s does not improve the subjects' ability to grade the images

Fig. 2.4 Structure diagram of test session for DSIS method

(2) Double-Stimulus Continuous Quality-Scale (DSCQS) Method

In this case, "double" not only implies that there are two presentations to the experimenter, but also means there are two responses returned from each experiment. It is the dissimilarity between the DSIS method mentioned above and the DSCQS method. The experimenters are asked to evaluate the image quality for each test condition by placing a mask on a vertical scale. It should be mentioned that the vertical scales are printed in pairs as shown in Fig. 2.5. For avoiding quantization errors, the vertical scales offer a continuous rating mechanism. Meanwhile, the vertical scales are divided into five equal lengths depicted in Fig. 2.5. The associated categorizing from bottom to top are Bad (0–19), Poor (20–39), Fair (40–59), Good (60–79), and Excellent (80–100), respectively. The scale provided to the subjects is the same as Fig. 2.5, except the figures at the right because the numbers are just employed to analysis.

The experimental structure diagram is shown in Fig. 2.6. It should be mentioned that the subjects are unaware of the type of images (reference or test) from experiment to experiment. However, the subjects are asked to vote two images. In other words, this way needs to evaluate two versions of each test images. There are reference image and one test image (which is impaired or unimpaired) contained in

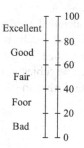

Fig. 2.5 Double-stimulus quality score

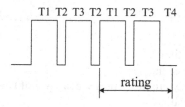

Phases of presentation:
T1=10s   Test image A
T2=3s   Mid-gray produced by a video level of around 200mV
T3=10s   Test image B
T4=5-11s Mid-gray

Fig. 2.6 Structure diagram of test session for DSCQS method

each pair images. In the experiments, the position of the reference image is uncertain. The advantage of this way is that the subjects do not always mark *A* or *B* at 100 which can improve the reliability of the experiments. The subjects are required to see the images twice and then are asked to rate on each image during the rating period at each experiment.

A typical experimental result is shown in Fig. 2.7. There are two purposes for us to employ the mean opinion score (MOS) values. On the one hand, we can employ the MOS to check the stability of the reference image responses. On the other hand, we can quickly obtain the dissimilarity between the reference and test images. It is must to mention that the experimental results are always accompanied by a statistical analysis. The position of the reference image is arbitrary in the process of assessing the image quality; it means that the score of the reference image is not always 100. Therefore, the dissimilarity between the reference image and test image is the most valuable data.

(3) Stimulus-Comparison Method

In this method, a subject is required to insert a mask on the continuous scale to reflect his or her inclination. The scale depicted in Fig. 2.8 contains three adjective masks ("*A* is much better", "*A* = *B*", and "*B* is much better"). The length of the scale is ten centimeters, and the numerical range of the scale is one hundred. The experimental structure diagram is shown in Fig. 2.9. In the experiment, the subjects are unaware of which is the reference image. Like other subjective image quality

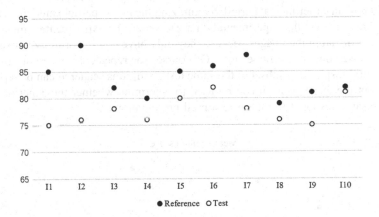

Fig. 2.7 Mean opinion score for DSCQS method

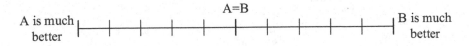

Fig. 2.8 Comparison rating scale

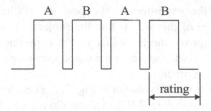

Fig. 2.9 Experimental structure for comparison method

assessment methods, this method also can be a double or single presentation. In the double presentation, the subjects are allowed to observe the images twice before rating.

(4) Single-Stimulus (SS) Methods

In single-stimulus methods, a single image is provided and the subjects offer an index of all presentation. The overall results of the SS methods are depicted in Fig. 2.10, where "y"-axis denotes the different numerical score ranged from −50 to 50 which corresponds to the adjectives employed in the experiments.

The only dissimilarity between the double stimulus and the single stimulus is the experimental structure. The diagram for single stimulus is shown in Fig. 2.11, where the subjects are asked to rate the imag/e at the $T3$ and $T4$. It must be mentioned that the reference and test images are blind to the participants in the experiments. The advantage of the single stimulus is that it allows the subjects to cover a variety of test conditions in an experimental period, maximizing the use of participants' time, as far as possible to collect the experimental data. However, the experimenters must keep more alert for rating the images because they only have one opportunity to examine the two conditions at the same time. Of course, for repeatedly observing the experimenters' response, a subset of the same test conditions scattered can be repeated throughout the experiment. It can be used to determine whether the experimenters have the same response to the same stimuli presentations.

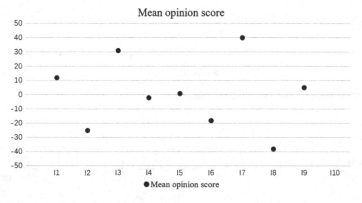

Fig. 2.10 Mean opinion score for SS method

Fig. 2.11 Single-stimulus
experimental structure

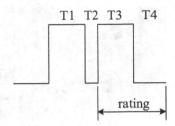

2.3   Public Image Quality Databases

For validation of the objective image quality assessment methods, there are more than twenty publicly available image quality databases that have been proposed at present. This section presents several well-known public databases that are annotated with reference and distorted images and subjective quality ratings. Detailed database special information, such as the brief description of the test images and the test conditions, is also provided in this section.

2.3.1   LIVE Image Quality Database

The entire Laboratory for Image and Video Engineering (LIVE) database (Sheikh 2006, 2014) presented by the University of Texas at Austin, USA, contains twenty-nine high-resolution and quality color reference images which were collected from the Internet and photographic CD-ROMs. Many conditions are included in the reference images, such as the picture of face, animals, people, nature scenes, image with different background configurations. A subset of source images is shown in Fig. 2.12. In addition, a part of images have high activity, while others are smooth. In order to better show the images on the screen resolution of 1024×768, the images are re-sized to various image resolutions ranging from 634×438 to 768×512. Moreover, 779 distorted images which were derived from the re-sized images also were contained in the database. There are five different distortion types which may appear in the real-world application used to generate the distorted images: JPEG2000 at bit rates ranging from 0.028 to 3.15 bits per pixel (bpp), JPEG at bit rates ranging from 0.15 to 3.34 bpp, white Gaussian noise of standard deviation σ_N (the values of σ_N employed ranging from 0.012 to 2.0), Gaussian blur (a circular-symmetric 2D Gaussian kernel of standard deviation σ_B ranging from 0.42 to 15 pixels employed to filter the R, G, and B channels), simulated fast fading Rayleigh (wireless) channel. Each type of distortion generates 5–6 distorted images, and these distortions would reflect the broad range of image degradations.

A single-stimulus method is employed in the experiments. It should be mentioned that the reference images are also hidden in the test images. There are seven

Fig. 2.12 Some source images in the LIVE database

sessions to conduct the experiments; the detail is shown in Table 2.1. All reference images randomly put on the distorted images in each session.

Most of the participant subjects are the undergraduate from the Digital Image and Video Processing and graduate from the Digital Signal Processing at the University of Texas at Austin. In addition, most of the subject pools are male students and they have no experience on image quality assessment. There are about 23 subjects to evaluate each image, as shown in Table 2.2. Moreover, each subject takes part in one session. The test images are displayed in a random order. The test subjects report their judgements of the image quality by dragging a slider on a quality scale which are labeled with: "Bad", "Poor", "Fair", "Good", "Excellent". The quality score is finally converted into DMOS in the interval [1 100] by a linearly mapping details of the processing mentioned in Sect. 2.2.

Table 2.1 Subjective evaluation sessions of LIVE database

Session	Numbers of images	Number of subjects
JPEG2000#1	116	29
JPEG2000#2	111	25
JPEG#1	116	20
JPEG#2	117	20
White noise	174	23
Gaussian blur	174	24
Fast fading	174	20
Total	982	22.8(average)
Alignment study	50	32

Note (1) The reference perfect images are contained in each session

Table 2.2 Types of distortion used in TID2008 database

No	Type of distortion	Correspondence to real situation	Accounted HVS peculiarities
1	Additive gaussian noise	Image acquisition	Adaptive, robustness
2	Additive noise in color components is more intensive than additive noise in the luminance component	Image acquisition	Color sensitivity
3	Spatially correlated noise	Digital photography	Spatial frequency sensitivity
4	Masked noise	Image compression, watermarking	Local contrast sensitivity
5	High frequency noise spatial	Image compression, watermarking	Frequency sensitivity
6	Impulse noise	Image acquisition	Robustness
7	Quantization noise	Image registration, gamma correction	Color, local contrast, spatial frequency
8	Gaussian blur	Image registration	Spatial frequency sensitivity
9	Image denoising	Image denoising	Spatial frequency, local contrast
10	JPEG compression	JPEG compression	Color, spatial frequency sensitivity
11	JPEG2000 compression	JPEG2000 compression	Spatial frequency sensitivity
12	JPEG transmission errors	Data transmission	Eccentricity
13	JPEG2000 transmission errors	Data transmission	Eccentricity
14	Noneccentricity pattern noise	Image compression, watermarking	Eccentricity
15	Local block-wise distortions of different intensity	In painting, image	Acquisition evenness of distortions
16	Mean shift (intensity shift)	Image acquisition	Light level sensitivity
17	Contrast change	Image acquisition, gamma correction	Light level, local contrast sensitivity

2.3.2 Tampere Image Quality (TID 2008) Database

This database, presented by the Tampere University of Technology, Finland, includes 1700 test images generated from 25 reference images which are obtained from the Kodak Lossless True Color Image Suite (Ponomarenko et al. 2009). A subset of the reference images is shown in Fig. 2.13. All images in the database are saved in 24-bpp BMP format with a size 512×384 pixel for the purpose of unification. There are 17 types of distortions for each reference image in the database, as shown in Table 2.3. Moreover, each type of distortions contains four different levels. With respect to most types of distortions, the relevant levels of PSNR are 21, 24, 27, and 30 dB (corresponding to bad quality, poor quality, good quality, and excellent quality).

Fig. 2.13 Some source images in the TID2008 database

A new methodology is used for TID2008 to carry out the subjective tests. In the experiment, the reference image, and a pair of distorted images are simultaneously presented in the screen, as an example in Fig. 2.14. Every subject is asked to select an image between two upper distorted images that is considered less different than the reference image below. After the selection, new distorted images will appear in the upper portion of screen. Each distorted image will obtain from 0 to 9 points during each experiment because each distorted image participates in 9 comparisons in each experiment (Ponomarenko et al. 2009).

In the end, more than 800 subjects with various cultural levels from three countries have involved in the experiments. In addition, 200 subjects conduct the experiments on the Internet, and others carry out the experiments in classes.

2.3.3   Other Public Image Quality Databases

(1) IRCCyN/IVC Image Quality Database (IVC)

The database (Ninassi et al. 2006; Le Callet & Autrusseau 2005) has been established by the Institute de Recherché en Communications et Cybernétique de Nantes (IRCCyN), France. In this database, there are a set of 10 reference images and 185 distorted images degraded by five types of distortions: 50 images distorted by JPEG compression, 25 images with luminance component distorted by JPEG compression, 50 images distorted by JPEG2000 compression, 40 images distorted by locally adaptive resolution coding, and 20 images distorted by Gaussian blurring. Moreover, each type of distortion contains four different levels. In addition, all images of dimension 512×384 pixels for the purpose of unification are stored in 24-bpp color BMP format. There are 15 test subjects rated the quality of the

Table 2.3 Majority of publicly available subjective image quality databases

Databases	Type	Year	Images number	Reference images	Distorted images	Distorted types	Resolution	Subjects
IRCCyN/IVC	Color	2005	195	10	185	5	512×512	15
LIVE	Color	2006	808	29	779	5	various	29
A57	Gray	2007	57	3	54	6	512×512	7
TID2008	Color	2008	1725	25	1700	17	384×512	838
MICT	Color	2008	196	14	168	2	768×512	16
IRCCyN/MICT	Color	2008	196	14	168	2	768×512	24
WID/Enrico	Gray	2007	105	5	100	10	512×512	16
WID/BA	Gray	2009	130	10	120	2	512×512	17
WID/FSB	Gray	2009	215	5	210	6	512×512	7
WID/MW	Gray	2009	132	12	120	2	512×512	14
WIQ	Gray	2009	87	7	80	1	512×512	30
CSIQ	Color	2010	896	30	866	6	512×512	35
IRCCyN/DIBR	Color	2011	96	3	96	3	1024×768	43
HTI	Color	2011	72	12	60	1	512×768	18
IBBI	Color	2011	72	12	60	1	321×481	18
DRIQ	Color	2012	104	26	78 enhanced	3 enhancements	512×512	9
TID2013	Color	2013	3025	25	3000	24	512×512	985

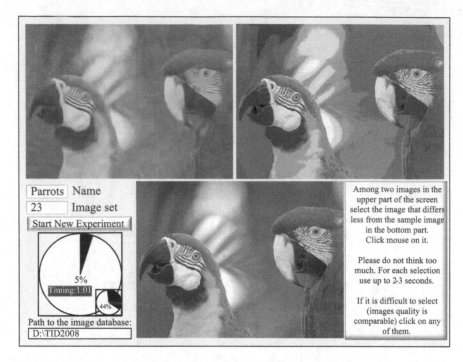

Fig. 2.14 Screenshot of the software used in TID2008 experiments

distorted images as compared to the reference images using DSIS in the experiments.

(2) A57 Image Quality Database

The database (Chandler et al. 2007; Le Callet & Autrusseau 2005) presented by the Cornell University, USA, includes 3 reference images and 54 distorted images degraded by six types of distortions. Moreover, all images of dimension 512×512 pixels for the purpose of unification are stored in 8-bpp gray BMP format. The six types of distortions in the database are: JPEG compression, JPEG2000 compression, uniform quantization of the LH sub-bands after a 5-level DWT (discrete wavelet transform), Gaussian blurring, Gaussian white noise, and custom JPEG2000 compression by the dynamic contrast-based quantization method. In addition, each type of distortion is generated with three different levels. Moreover, seven test subjects take part in the experiments.

(3) Toyama Image Quality (MICT) Database

The database presented by the Media Information and Communication Technology Laboratory of the University of Toyama, Japan, includes 14 reference images and 168 distorted images degraded by two types of distortions (Horita et al. 2018). In addition, the images are stored in 24-bpp color BMP format. The widths of images in the ranges 480–768, moreover, the ranges of heights are 488–720.

There two types of distortions in the database: 84 images distorted by JPEG compression and 84 images distorted by JPEG2000 compression. In addition, each type of distortion is generated with seven different levels. There are 16 test subjects rating the quality of the distorted images employing the adjectival categorical judgement method.

(4) IRCCyN Watermarking Database

Four different watermarking databases have been made available by the Institut de Recherche en Communications et Cybernétique Nantes (IRCCyN), France. The images are generated by the different watermarking algorithms which are Enrico, Broken Arrows (BA), Fourier Subband (FSB), and Meerwald (MW), respectively.

(a) IRCCyN Watermarking-Enrico Database (WID/Enrico).

The database (Marini et al. 2007) includes 5 reference images and 100 distorted images. Moreover, the distorted images of dimension 512×512 pixels generated from 10 different watermarking algorithms are stored in 8-bpp gray BMP format. There are 16 test subjects rating the quality of the test images in the experiments.

(b) IRCCyN Watermarking-Broken Arrows Database (WID/BA).

The database (Autrusseau & Bas, 2009) includes 10 reference images and 120 distorted images. In addition, the distorted images of dimension 512×512 pixels contain watermarking at 6 different embedding strengths. The images are stored in 8-bpp gray PPM/PGM format. The experimental results are obtained from 17 test subjects.

(c) IRCCyN Watermarking-Fourier Subband Database (WID/FSB).

The database (Carosi et al. 2010) includes 5 reference images and 210 distorted images. Furthermore, the distorted images that include watermarking in 6 frequencies at 7 different embedding strengths in the database are stored in 8-bpp gray BMP format at the resolution of 512×512 pixels for the purpose of unification. The quality scores of the distorted images are determined by the seven test subjects.

(d) IRCCyN Watermarking-Meerwald Database (WID/WM).

The database (Autrusseau & Bas, 2009) includes 12 reference images and 120 distorted images. Besides, the distorted images generated from 5 different embedding watermarking are stored in 8-bpp gray BMP format at the resolution of 512×512 pixels. The ratings were obtained from the 14 test subjects.

(5) Wireless Image Quality (WIQ) Database

The database (Engelke et al. 2009, 2010), which has been made available by the Radio Communication Group at the Blekinge Institute of Technology, Sweden, includes 7 reference images and 80 distorted images distorted by loss of JPEG data over a stimulated wireless channel. In addition, the images of dimension

512×512 pixels are stored in 8-bpp gray BMP format. There are 30 test subjects rating the quality of the distorted images in two separated experiments.

(6) Categorical Subjective Image Quality (CSIQ) Database

This database (Larson et al. 2014) presented by the Oklahoma State University, USA, includes a set of 30 reference images and 866 distorted images degraded by six types of distortions. Furthermore, the images are stored in 24-bpp color PNG format at a resolution of 512×512 pixels. The six types of distortions in the database are: 150 images distorted by JPEG compression, 150 images distorted by JPEG2000 compression, 150 images distorted by additive Gaussian white noise, 150 images distorted by additive Gaussian pink noise, 150 images distorted by Gaussian blurring, and 116 images distorted by global contrast decrements. Moreover, each type of distortion contains 4–5 different levels. There are 35 test subjects participating in the subjective ratings.

(7) TU Delft Perceived Ringing (TUD1 and TUD2) Databases

The databases (Liu et al. 2013) have been available by the Delft University of Technology, the Netherlands. There are two trials used to rate the distorted images. In the first experiment, 12 test subjects evaluate 16 JPEG-compressed images produced from 8 reference images with two different levels of compression. In the second experiment, 44 JPEG-compressed images produced from 11 reference images with four different levels of compression are assessed by 20 test subjects.

(8) VCL@FER Image Quality Database

The database (Zaric et al. 2011) presented by the University of Zagreb, Croatia, includes 23 reference images and 552 distorted images degraded by four types of distortions: 148 images distorted by JPEG compression, 148 images distorted by JPEG2000 compression, 148 images distorted by additive Gaussian white noise, and 148 images distorted by Gaussian blurring. Moreover, each type of distortion contains six different levels. There are 118 test subjects participating in the subjective ratings.

(9) Digitally Retouched Image Quality (DRIQ) Database

This database (Vu et al. 2012), which has been available by the Oklahoma State University, USA, includes 26 reference images and 78 enhanced images degraded by manual digital retouching. Furthermore, the images are stored in 24-bpp color PNG format at a resolution of 512×512 pixels. There are 35 test subjects participating in the subjective ratings.

A description and summarization of the test material within each database, such as the type of images, the number of reference and distorted images, the number of distorted types, resolution and format, are offered in Table 2.3. In addition, numbers of image quality databases are the part of the multimedia databases provided by the QUALINET consortium (European Network on Quality of Experience in Multimedia System and Services) (Chandler 2013). In (Winkler 2012), and an analysis of the public image and video quality databases for quality assessment is presented. Such analysis compares some aspects of the databases, such as source content, test material, and subjective ratings.

2.4 Conclusion and Future Work

In subjective image quality assessment, human experimenters are employed to rate the image quality in various methods. Therefore, it is the most reliable method for evaluating the quality of images because the human is the ultimate receivers for images. The development of the subjective quality assessment can be dated back to a long time ago. In addition, this field is still in continuous research by numbers of studies. Numbers of subjective image quality assessment databases have been proposed in the past few decades. Furthermore, some different ways for designing and defining the subjective evaluation tests are also presented since the image quality getting more and more attention. However, there still are some challenges to the existing subjective image quality approaches.

With the rapid development of the science and technology, image techniques also get a lot ascension. Meanwhile, in recent years, ultra-high-resolution display offers an immersive virtual reality visual perception with the development of the new display technologies. For example, high dynamic range displays can offer a more immersive visual experience by magnifying perceived contrast. Furthermore, the perceived depth information is enhanced with the development of stereoscopic and autostereoscopic displays (3D) technologies (Lambooij et al. 2011; Seuntiens et al. 2006). Several 3D image quality databases have been proposed at present. However, for the proposed 3D image quality databases, the number of images and distortion types are too few to show the reliability of the databases. In addition, the optimization of image acquisition and display techniques is still continuous. For driving this optimization, there are several essential things about image quality assessment we need to do. Firstly, it is important to properly understand the impact of multidimensional information on the image quality evaluation. Secondly, a new adaptive subjective image quality assessment method which can adapt the development of the new technology should be exploited.

A second significant challenge for the subjective image quality assessment is how to contain the aesthetic evaluations and emotion evaluations into quality measurements. The experience, personal opinions, cognitive bias, and memories can strongly affect the appreciation of aesthetics. It may lead to the difference of the image quality score for everyone because the aesthetics can impact the viewing experience. Furthermore, the emotion states of the subjects may affect the viewing experience. In reverse, the viewing experience may impact the affective state. The image can create an empathic experience with the image content by inducing emotion. Therefore, the potential influence of the emotion states should be considered in image quality assessment methods. Efforts are currently growing, and the affective computing community has been proposed in this direction (Deng et al. 2015). At the same time, a valuable tool used to control the user's emotions prior to the viewing experience may be employed in the future. However, major challenges are ahead.

For acquiring the reliable results, the subjective experiments are usually conducted in highly controlled, standardized environments. The popular several subjective image quality assessment approaches are all performed within laboratory facilities. This makes it possible to control the environments, such as lighting and viewing position, and make visibility conditions homogeneous for all subjects. It leads to the results of the experiment more reliable. However, with the advent of the mobile techniques (phones and tablet PC) and Internet-based image presentation, the viewing feel is consumed in greatly various environments at present. This requires a change in the traditional laboratory-based subjective measurements. In addition, more attention should be paid to the composition of the test subjects with the acceptance of considering and understanding the personal differences of subjective image quality study (Deng et al. 2015). The existent differences in terms of consumer affecting should be represented as far as possible. The subjects should be as representative as possible. This also makes it unfeasible to perform the experiments in the laboratories. In this scenario, Crowdsourcing for subjective tests of image quality assessment has been proposed and receives more and more attention in recent years (Wu et al. 2013; Xu et al. 2012). The main idea of the Crowdsourcing is to outsource small and repetitive tasks to numbers of online people and pay a small compensation. However, there are still many challenges in Crowdsourcing. The scheme of payment, choice of workers, and accuracy of evaluation are to be determined.

References

Autrusseau, F., & Bas, P. (2009). Subjective quality assessment of the broken arrows watermarking technique. http://www.irccyn.ec-nantes.fr/~autrusse/Databases/BrokenArrows/.

Carosi, M., Pankajakshan, V., & Autrusseau, F. (2010). Toward a simplified perceptual quality metric for watermarking application. In *Conference on Multimedia on Mobile Devices*. San Jose.

Chandler, D. M. (2013). Seven challenges in image quality assessment: Past, present, and future research. *ISRN Signal Processing, 2013*(8), 905685.

Chandler, D. M., & Hemami, S. S. (2007). VSNR: A wavelet-based visual signal-to-noise ratio for natural image. *IEEE Transactions on Image Processing, 16*(9), 2284–2298.

Choe, J. H., Jeong, T. U., Choi, H. S., Lee, E. J., & Lee, S. W. (1999). Subjective video quality assessment methods for multimedia applications. *ITU-T Recommendation P. 910, 12*(2), 3665–3673.

Deng, C., Ma, L., Lin, W., & Ngan, K. N. (2015). *Visual quality assessment*. Switzerland.

Engelke, U., Maeder, A. J., & Zepernick, H. J. (2009). *Visual attention for image quality database*. http://www.bth.se/tek/rcg.nsf/pages/vaiq-db.

Engelke, U., Zepernick, H. J., & Kusuma, M. (2010). *Wireless imaging quality (WIQ) database*. http://www.bth.se/tek/rcg.nsf/pages/wiq-db.

Horita, H., Shibata, K., & Kawayoka, Y. (2018). *Toyama image quality evaluation database*. http://mict.eng.u-toyama.ac.jp/mict/index2.html.

ITU-R B.T. (2012). *Methodology for the subjective assessment of the quality television pictures.* Geneva, Switzerland: International Telecommunication Union. ITU-R Recommendation BT.500–13.

Lambooij, M., IJsselsteijn, W., Bouwhuis, D. G., & Heynderickx, I. (2011). Evaluation of stereoscopic images: beyond 2d quality. *IEEE Transactions on Broadcasting, 57*(2), 432–444.

Larson, E. C., & Chandler, D. M. (2014). *Categorical image quality (CSIQ) database.* http://vision.okstate.edu/csiq.

Le Callet, P., & Autrusseau, F. (2015). Subjective quality assessment irccyn/ivc database. http://www2.irccyn.ec-nantes.fr/ivcdb/.

Liu, H., Klomp, N., & Heynderickx, I. (2013). *TUD image quality database: Perceived ringing.* http://mmi.tudelft.nl/iqlab/ringing.html.

Marini, E., Autrusseau, F., Le Callet, P., & Campisi, P. (2007). Evaluation of standard watermarking techniques. In *Proceedings of the International Social for Optical and Photonics* (p. 6505).

Ninassi, A., Le Callet, P., & Autrusseau, F. (2006, January). Pseudo no reference image quality metric using perceptual data hiding. In *Conference on Human Vision and Electronic Imaging XI*. San Jose.

Ponomarenko, N., Lukin, V., & Zelensky, A. (2009). TID2008-a database for evaluation of full-reference visual quality assessment metrics. *Advances of Modern Radioelectronics, 10*, 30–45.

Seuntiens, P., Meesters, L., & IJsselsteijn, W. (2006). Perceived quality of compressed stereoscopic images: Effects of symmetric and asymmetric jpeg coding and camera separation. *ACM Transactions on Applied Perception, 3*(2), 95–109.

Shahid, M., Rossholm, A., Lövström, B., & Zepernick, H. J. (2014). No-reference image and video quality assessment: A classification and review of recent approaches. *EURASIP Journal on Image and Video Processing, 2014*(40), 1–32.

Sheikh, H. R., Sabir, M. F., & Bovik, A. C. (2006). A statistical evaluation of recent full reference image quality assessment algorithms. *IEEE Transactions on Image Processing, 15*(11), 3440–3451.

Sheikh, H. R., Wang, Z., Cormack, L., & Bovik, A. C. (2014). *LIVE image quality assessment database release 2.* http://live.ece.utexas.edu/research/Quality/.

Vu, C., Phan, T., Singh, P., & Chandler, D. M. (2012). *Digitally retouched image quality (DRIQ) database.* http://vision.okstate.edu/driq.

Winkler, S. (2012). Analysis of public image and video databases for quality assessment. *IEEE Journal of Selected Topics in Signal Processing, 6*(6), 616–625.

Wu, H. R., & Rao, K. R. (2006). Digital video image quality and perceptual coding. *Journal of Electronic Imaging, 16*(6), 039901.

Wu, C. C., Chen, K. T., Chang, Y. C., & Lei, C. L. (2013). Crowdsourcing multimedia QoE evaluation: A trusted framework. *IEEE Transactions on Multimedia, 15*(5), 1121–1137.

Xu, Q., Huang, Q., & Yao, Y. (2012). Online crowdsourcing subjective image quality assessment. *ACM International Conference on Multimedia*, 359–368.

Zaric, A., Tatalovic, N., Brajkovic, N., Hlevnjak, H., Loncaric, M., Dumic, E., & Grgic, S. (2011, September). Vcl@fer image quality assessment database. *Proceeding of the International Symposium ELMAR*. Zadar, Croatia.

Chapter 3
Human Visual System and Vision Modeling

3.1 Introduction

For humans, the information perceived through vision takes up about 80% of the total information that is obtained from the outer world. Therefore, it is important to have a precise understanding of the mechanism underlying human visual system (HVS). Since the HVS is the ultimate receiver of images, the ideal way to evaluate the quality of images is the subjective image quality assessment (IQA) methods given by human observers, according to their perceptual quality. Though this kind of methods can always reflect the perceptual quality of images veritably, they are time-consuming, laborious, inflexible, and unstable, which makes it impracticable for them to be applied in real-time systems. Researchers mainly focus on constructing IQA algorithm to predict the quality automatically. Meanwhile, the objective scores should be consistent with subjective scores.

There are some traditional metrics measuring the quality of images, such as methods based on signal fidelity, methods using MSE (mean squared error), SNR (signal-to-noise ratio), and PSNR (peak signal-to-noise ratio). However, in Lin and Kuo (2011), the authors put forward four problems on IQA methods and prove that simply calculating pixel-wise distances is not enough. Actually, the contribution of distortion made by every pixel is nonuniformed and determined by the properties of human perception. Therefore, the ideal methods for IQA are supposed to take the characteristics of HVS into account.

As mentioned in Chandler (2013), the earliest efforts for modeling properties of HVS for IQA include the researches of Sakrison and Algazi (1971), Budrikis (1972), Stockham (1972), and Mannos and Sakrison (1974). In these works, the luminance, contrast sensitivity, and visual masking effects are introduced into modern IQA methods. In Lin et al. (2003), Lin et al. took the just notice difference into account, and in Vu et al. (2008), the authors paid their concentration on the visual fixation patterns to measure the quality of images. Moreover, with the insight to the HVS, Gu et al. (2015) proposed an IQA metric with the free-energy-based

Y. Ding, *Visual Quality Assessment for Natural and Medical Image*,
https://doi.org/10.1007/978-3-662-56497-4_3

brain theory and classic HVS-inspired features. In consequence, the IQA metrics based on HVS runs through the development of image quality assessment. With the good performance of these metrics, it is believed that a good understanding of the anatomy structure and properties of HVS can be helpful to evaluate the image quality.

The aim of this chapter is to give a brief presentation of HVS, though the knowledge of HVS is far from complete. Firstly, from the anatomy and physiology, we provide an introduction to the structure of HVS in Sect. 3.2. Then, the simple properties of HVS applied in current implementations of image quality evaluation are outlined in Sect. 3.3. For detailed information of HVS, readers can refer to more dedicated literature (Geisler and Banks 1995; Wandell 1995; Cormack 2005).

3.2   Anatomy Structure of HVS

The simplified anatomy structure of HVS is shown in Fig. 3.1. As mentioned in Wang and Bovik (2006), the processing system includes four stages: (1) optical processing, (2) retinal processing, (3) LGN (lateral geniculate nucleus) processing, and (4) cortical processing.

As shown in Fig. 3.1, the first stage is a path of light through eyes. In human eyes, the light passes through the cornea, the pupil, and the lens, successively. Finally, the light focuses on retina, which is in the back of the eye and where the light-sensitive photoreceptors are transformed into nerve stimuli. In the processing,

Fig. 3.1 Simplified anatomy
structure of HVS

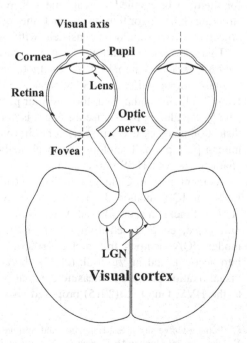

the light is changed via lens and many other transmission media, e.g., the cornea, the aqueous humor, the gelatinous vitreous humor. In Moorthy et al. (2011), the authors deem the processing band-limited and consider it to be a low-pass filter, which can be roughly modeled by point spread function (PSF).

The retina has several layers of neurons. And it consists of many different types of neurons, including the rod and cone photoreceptors, the bipolar cell, the amacrine cell, the horizontal cell, and the ganglion cell, etc. The retina processing begins with the photoreceptors that take the information from light, and then the information is transmitted to bipolar cells and amacrine cells to have encoding of the light. Finally, the ganglion cells, whose axons come into being the optic nerve, output the signal from the eye as shown in Fig. 3.1.

Photoreceptors have two classes, rods and cones. It proves that rod photoreceptors are the second most numerous neurons in humans and only less than the cerebellar granule cells (Masland 2012). The rod photoreceptors work in low-light conditions, and they are sensitive to luminance. Since the spectral sensitivity of all rods is roughly same with the peaks at 500 nm, the rod photoreceptors could not preserve the color information. Different from the rods, the cone photoreceptors are used in high-light conditions and they are the basis of the color vision. According to different sensitivity to wavelengths of light, there are three types of cone photoreceptors: the short wavelength-sensitive cones (S-cones), the middle wavelength-sensitive cones (M-cones) and the long wavelength-sensitive cones (L-cones). Since the cone photoreceptors respond to distinct wavelength light, the light inside is encoded with the S-cones (410–435 nm), M-cones (530–550 nm), and L-cones (556–562 nm). In fact, this is the principle of trichromatic color vision (Ouria et al. 2006).

The output of the photoreceptors mainly passes through horizontal cells, bipolar cells, and amacrine cells. The functions of these cells are distinct (Masland 2012). The function of horizontal cells is controlling the local gain of the retina. With gap junctions, the adjacent horizontal cells are connected to measure the average level of illumination from a local region of outside retinal surface. In order to hold a controllable range, these cells remove a proportionate value from the average value in a local region. Bipolar cells have different types which transmit different outcoming signals that can reflect various features. This information is simple and elementary. As an example, a blue cone bipolar simulated by the signal from photoreceptors changes the ganglion cell into a blue-ON, green-OFF one. Then, these features are connected to form the signal that is transmitted to the ganglion cells, eventually to the inner retina. Amacrine cells are supposed to help the bipolar cells have a refined signal transmission for specific ganglion cell subtypes.

The information from bipolar cells and amacrine cells is transmitted out of the eye by the optic nerve which is formed by axons of ganglion cells, as shown in Fig. 3.1. Since the types of upper cells are varied, there are many types of ganglion cells. The overall number of rods is about 120 million, and cones are about 7 million. Then, the information is converged by 1 million ganglion cells. However, the distribution of ganglion cells is not uniform. In the central area, there is a region in elliptical shape called fovea where ganglion cells have concentrative distribution. Meanwhile, the density of cones increases with the decrease of distance from the enter of fovea where is a rod-free region. In consequence, the cone receptors have one-to-one contact with ganglion cells in this region, and there are about 100 receptors that share 1 ganglion cell in surround areas (Wu and Rao 2006). This leads to the fact that in high-light condition, human visual system gives us the color-sensitive and exquisite sharpen vision and the two properties are under-performance for scotopic vision. With the distance from the fovea increasing, the resolution of the perceived image decreases which is called foveated vision. When we look at the picture shown in Fig. 3.2a, we focus our attention on the man sitting shoreside and the generally perceptible image we have is shown in Fig. 3.2b with the influence of the nonuniform distribution. The area near the fixation point in the picture with high resolution is called the foveal vision and the resolution of vision declines with the distance between the vision and the fixation, the resolution of the vision become lower. Foveae vision can help us to find the important information and saliency regions in an image.

In visual system, receptive field is the basic functional unit. Take the ganglion cells of receptive field as an example. In Kuffer (1953), Kuffler have a research on the ganglion cells of mammalian retina and find that there are two types of ganglion cells in the receptive field, as shown in Fig. 3.3. One is in the central of a concentric ring that is sensitive to the light. The other surrounding one responds to the light adversely.

Fig. 3.2 Example of foveae vision

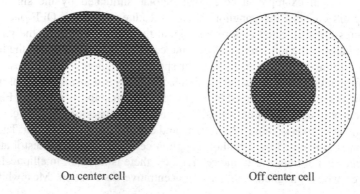

On center cell   Off center cell

Fig. 3.3 Schematic diagram of the receptive field

The properties of the receptive field are described as the center-surround antagonistic. With the center-surround antagonistic, the edge of object can be detected.

So what are the functions of retina? In general, light adaption, automatic gain control, image sharpening (center-surround antagonism), and the cable to visual cortex can be regarded as the basic functions of retina. However, with the development of anatomy and physiology, it is confirmed that there are more functions of retina, such as motion detection, motion discrimination, latency coding. For more detailed information, refer to Gollisch and Meister (2010).

The nerve stimulation is transmitted to LGN through optic nerve, and LGN serves as a bridge between the retinal and the visual cortex. As shown in Fig. 3.1, the information encoded from left and right retina contains the color, shape, motion, and other basic information is merged in LGN, but it is still monocular. The receptive field of the LGN is similar to that of ganglion cells. It is modeled as the center-surround antagonism which has stronger effect than ganglion cells.

The visual cortex processes the output of LGN. It is acknowledged that in the primate brain, there is around 55% neocortex that is concerned with visual perception (Felleman and Essen 1991). In Kruger et al. (2013), the visual cortex is divided into three parts, including occipital part, ventral pathway, and dorsal pathway. Occipital part that contains V1–V4 and MT (middle temporal) is known as early vision. With the increase of the layer depth, the perception of vision becomes more complex. Functional processes established in V3, V4, and MT concentrate on integrating lower level features into higher level responses such as the simple shape formed, the visual motion, and the depth processing. Meanwhile, these three areas increase the invariances of the extracted information, such as color hue. The ventral pathway having the information transmitted from occipital part (as shown in Fig. 3.4) is engaged in object recognition and categorization with the TEO and TE areas. The dorsal pathway devotes to space analysis and action planning with medical superior temporal (MST) area for visual motion and visual areas in posterior parietal cortex as shown in Fig. 3.4. This is a simple model about the connection in different layers, and more detailed information is presented in Kruger et al. (2013). However, since the image quality assessment is regarded as a processing in the early vision, the studies on visual cortex concentrate on V1 and V2 in IQA.

Fig. 3.4 Simplified structure of visual cortex

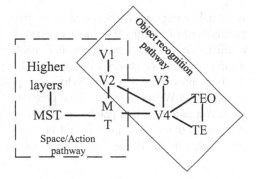

The visual information from LGN is firstly transmitted to V1 in cortical area. The cells in V1 area can be divided into two types, simple cells and complex cells. The former ones are sensitive to phase and position, while the latter ones lost this sensitivities and have larger receptive field than the former (Orban 2008). In Jones and Palmer (1987), it is proved that Gabor wavelets can model the simple cells. Meanwhile, the response of the complex cells is similar to the magnitude of the quadrature pair of Gabor. This is the reason for the widespread use of Gabor in IQA. Area V1 can extract many basic features, including edges, motion, disparity, and color. Direction selection demonstrates that the cells in V1 area only respond to certain moving, which is the principle for motion detected. As the V1 can receive the information from both right and left eyes, the disparity between them can be obtained, which contributes to the formation of three-dimensional vision. Stereoscopic vision is produced from a whole field of human visual system. About 5–10% cells including single opponent and double opponent in V1 have the exclusive use of color-coding (Conway 2009; Shapley and Hawken 2011).

It is a point-to-point reception from area V1 to area V2. Although V2 cells can reserve the color, orientation, and disparity features as V1, the sophisticated contour features can be formed in V2. Particularly, V2 cells are more sensitive to relative disparity instead of absolute disparity. Studies in neurophysiology show that there are many types of cells to realize the different functions in V2, such as the cells with orientation tuning that can respond to texture-defined contours. The main contribution of V2 is the representation of contours.

The knowledge of HVS can contribute to the development of IQA. In image quality assessment, there are many researches (Xue et al. 2014; Gao et al. 2009; Wu et al. 2013) that employ the texture information such as edge, gradient, and local pattern, to simulate the images transformed in HVS. Then, they use this information to measure the quality of distorted images. The results of these researches show that it is meaningful to have a sufficient understanding of HVS.

3.3 Basic Properties of HVS

3.3.1 Light Adaption

In visual perception, HVS is more sensitive to the relative differences between the background of illumination intensity and the changes of intensity. And there exists a difference threshold called just-noticeable difference (JND), which refers to the minimum intensity changes that can be detected by HVS. The concept of JND is firstly proposed to solve the weight lifting problem by Ernst Heinrich Weber. When we apply the Weber law into visual perception (Wu and Rao 2006), we can imagine there is a patch. In this patch, the intensity of background is I, and the minimum differences that can be just noticed between the sounding and the target object are represented as ΔI. In fact, the Weber law intends to measure the ratio of JND

$$\frac{\Delta I}{I} = K \qquad (3.1)$$

where K is the value ratio. In Hecht (1924), Schreiber (1986), the value of K is about 0.02, when the intensity of background I is in the middle range. In addition, experiments (Shen and Wang 1996) show that the value of K increases rapidly when the intensity of background is getting too high or too low, while the sensitivity of the target detection reduces. That is to say, the property of JND is similar to a band-pass filter, and the value of JND is not only determined by the intensity changes between the surroundings and targets, but also depends on the background intensity. Moreover, this is corresponding to the light adaptation property of retina in HVS mentioned above. Lower intensity refers to the low-light circumstances in retina. In this case, the functions of receptors are restrained which makes the detected threshold increase. The higher ranges of the intensity I correspond to the high-light conditions in retina. And the local gain control operates to prevent HVS from too much intensity leading to the inadequate responses for intensity changes.

There are many researches (Watson 1993; Lin et al. 2003) employing JND to the image quality assessment system for having better matches with HVS perception. After JND processing, the metrics reserve the noticeable distortions and remove the influence of the useless information.

3.3.2 Contrast Sensitivity Function

Contrast sensitivity function reflects the relationship between the sensitivity of HVS and the diverse spatial frequency. Michelson contrast is one of the popular definitions of contrast:

$$\text{Contrast} = \frac{L_{\max} - L_{\min}}{L_{\max} + L_{\min}} \qquad (3.2)$$

where $L_{\max}$ denotes the peak luminance of grating and $L_{\min}$ refers to the trough.

Schade (1956) firstly conducted an experiment on the contrast with sine-wave gratings to find the property of contrast relative to spatial frequency. In general, contrast threshold is the minimum luminance differences noticed by HVS. The results of the experiment show that the spatial frequency of the target influences the ranges of contrast threshold. Since contrast sensitivity is the inverse of contrast threshold, contrast sensitivity is closely relative to the spatial frequency. And the contrast sensitivity function is defined to describe the relationship between contrast sensitivity and spatial frequency of target.

As shown in Fig. 3.5a, the abscissa in the chart is the spatial frequency with the value increasing from left to right while the ordinate is contrast with the value decreasing from up to bottom (Campbell and Robson 1968). From the chart, it is visualized that the profile of CSF is with the shape of a bell upended and the most

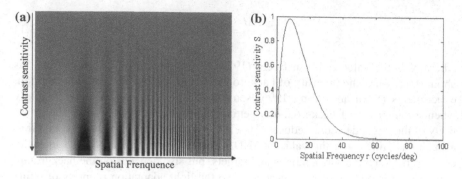

Fig. 3.5 **a** Campbell-Robson CSF chart; **b** curve of CSF (Mannos-Sakrison filter) in frequency domain

sensitive frequency is in the middle range. Moreover, in order to incorporate this property of HVS into image quality assessment, some spatial filters are designed to imitate the CSF. Since the behaves of CSF are like a band-pass filter with less sensitive to low and high frequency, researchers have proposed some mathematical formulas to describe this function, including Mannos-Sakrison filter (Mannos and Sakrison 1974), Daly filter (Daly 1992), and Ahumada filter (Ahumanda 1996; Watson and Ahumanda 2005). In this chapter, we introduce a popular method, Mannos-Sakrison filter.

$$S(f) = 2.6(0.0192 + 0.114f)e^{-(0.114f)^{1.1}} \qquad (3.3)$$

In this model, f is the spatial frequency and S represents the contrast sensitivity. The curve of Eq. (3.3) is shown in Fig. 3.5b. Although the model devotes to imitating CSF, the property can be found below. As shown in Fig. 3.5b, the peak of the contrast sensitivity S is about 8 cycles/degree and is decreasing as the value of spatial frequencies is getting lower or higher.

What are the reasons for the reduction in sensitivity? For over-high frequency, it mainly results from the optics of eye, the spacing of receptors, and the quantum noise; for over-low frequency, it is attributed to the receptive field sizes of cortex cells and the masking effects.

The explorations on contrast are important in image processing. However, in visual sciences, there are many factors that have influence on contrast sensitivity, such as luminance, spatial frequency, and viewing angle. In fact, JND is an example of how luminance effects contrast sensitivity. The sensitivity reduces when the luminance is beyond the suitable range. When applying these properties into image quality assessment, one of the typical ways is to regard them as prefilters to throw off the useless information.

3.3.3 *Visual Attention Mechanism*

In complex nature scenes, it is a hard work for HVS to process the plenty of visual information. Visual attention is an essential mechanism for HVS to achieve information selection and down-sampling. In this procedure, with the useless information eliminated, visual attention mechanism makes people pay more attention to more attractive regions that contain more important and interesting information. In recent years, researchers have been trying to take visual attention mechanism into consideration in many fields like image segmentation, objection detection, image compression, image quality assessment. In the studies of the visual perception, there are two types of visual attention mechanisms (Chen et al. 2016), namely bottom-up mechanism and top-down mechanism.

The bottom-up saliency map is automatically created in the primary visual cortex (Li 2002) without the guidance of target detection and the prior information of context. In addition, this saliency map focuses on color, orientation, intensity, texture, and other primer visual features. As shown in Fig. 3.6, the observers tend to be sensitive to the distinctive regions that are more conspicuous. A classic and important bottom-up saliency model proposed in Itti et al. (1998), Itti and Koch (2000) is based on the hypothesis that the saliency map can derive from the basic feature maps that can deploy attention on a single bottom-up cues. Itti et al. decomposed the input image into three channels, i.e., color, intensity, and orientation channels, to obtain the saliency maps in each channel with the center-surround mechanism. Finally, the overall saliency map is composed of these feature maps by linear combination.

The top-down saliency map detection (Oliva 2005; Goferman et al. 2012; Tatler et al. 2010) is guided under context, prior knowledge, and target detection mask, which is associated with processing procedure in high-level visual system with the assistance of memory and consciousness. With the development of the study in eye tracking, it is proved that the visual attention has a close relationship with visual subjective expectations. CA saliency detection model (Goferman et al. 2012) detects the saliency regions of the scene with four principles (Koffka 1955; Treisman and Gelade 1980; Wolfe 1994; Koch and Poggio 1999): the two considerations of local low-level considerations (contrast, color, and so on), global considerations, the rules of visual organization, and high-level factors containing objection detection and object location.

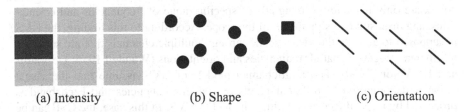

(a) Intensity   (b) Shape   (c) Orientation

Fig. 3.6 Three examples of primer visual features

Most of the bottom-up saliency models are closely linked to the complexity of regions with the measurement of the variances of orientations (Yamada and Cottrell 1995), entropy (Kadir and Brady 2001), and other primer features. And there is a widely accepted hypothesis says that the complexity measures the richness and diversity in a local region. The local regions that contain many different orientations tend to be more salient. However, this hypothesis cannot apply to this situation: an egg in a nest (egg should be salient with less complexity). In addition, most of the top-down saliency models are time-consuming and failing to process with details. In consequence, there are many studies (Zhang et al. 2008; Levin and Weiss 2009; Navalpakkam et al. 2010) which combine the bottom-up and top-down methods to build the complete saliency map. In Zhang et al. (2008), the authors believe that the goal of attention system is to find the potential targets. A part of the metric is named self-information with the bottom-up saliency. The top-down saliency map is modeled with a log-likelihood, Bayesian formulation. Finally, the overall saliency map is obtained by combining the two parts using the pointwise mutual information.

The goal of image quality assessment is to measure and quantify the distortion. In full-reference IQA, as mentioned before, it is unfaithful to simply measure the differences between reference images and distorted images without taking characteristics of HVS into consideration. In Tong et al. (2010), the local quality maps of images are weighted with the saliency map to emphasize the differences between the original and distorted images in more conspicuous regions. This is based on the assumption that quality degradation occurred in more salient regions is always more annoying than that in unnoticed regions. Besides, this metric also employs saliency models proposed in Itti et al. (1998), Itti and Koch (2000). In Ding et al. (2017), the local regions are weighted with saliency map calculated by method proposed in Zhang et al. (2008) to measure the distortion of images. In past decade, the saliency map has been extensively utilized in IQA algorithms (Min et al. 2014; Zhang et al. 2014; Kottayil et al. 2016; Saha and Wu 2016; Alaei et al. 2017).

3.3.4 Multichannel Properties

Studies in psychophysics and physiology have revealed that the receptive fields of neurons in the primary visual cortex are size specific and tuned to signals with specific spatial locations, frequencies, and orientations. That is, neurons in visual cortex are only responding to signals of specific range of frequencies and orientations, and insensitive to signals out of the scope. According to this finding, the HVS is supposed to analyze the visual scene through multiple channels that are sensitive to different ranges of spatial frequencies and orientations (Wandell 1995; De Valois and De Valois 1990). The multichannel models of HVS assume that the visual scene is decomposed to different bands of spatial frequencies and each band is processed by a specific corresponding channel in HVS. In this case, the CSF can be regarded as the envelope of the sensitivities of these channels. Detection occurs independently in any channel without being disturbed by activities of other channels.

Inspired by this property of HVS, researchers have been trying to use band-pass and oriented filters to decompose an image signal into multiple channels, and some of the decomposition models have been employed in image quality assessment, such as Fourier decomposition (Mannos and Sakrison 1974), Gabor decomposition (Taylor et al. 1997), local block DCT transform (Watson et al. 2001), and separable wavelet transform (Watson et al. 1997). The oriented wavelet-type filters always better describe the receptive fields of the neurons in primary visual cortex. This multichannel property can be mathematically regarded as the function of multiresolution filters or wavelet decomposition (Daubechies and Sweldens 1998). In De Valois and De Valois (1990), the authors point out that there is a continuous distribution of peak frequencies in cortical cells, though multiresolution models have been built using a discrete number of spatial channels and orientations (Wilson and Regan 1984). Physiological and psychophysical findings indicate that the bandwidths of the channels are broader for lower frequency channels while narrower at higher frequencies in the logarithmic coordinate. In linear coordinate, however, the low-frequency channels have narrower bandwidth than the high-frequency channels.

An early full-reference multichannel IQA algorithm called the visible difference predictor (VDP) is proposed in Daly (1992). Several processing stages are involved in this model, including an amplitude nonlinearity, an orientation-dependent CSF filtering, a channel decomposition, a contrast calculation, and a simple intra-channel masking function. The VDP is used to evaluate the image fidelity of the distorted image compared with the reference image, and output a probability-of-detection map between the two images. Similarly, Lubin also proposed an algorithm to estimate the probability of detection of the difference between the reference and distorted images (Lubin 1995). A perceptual distortion metric for digital color video is proposed in Winkler (1999), which introduced a spatiotemporal decomposition and generated a number of perceptual channels during the processing.

3.3.5 Visual Masking

It is widely acknowledged that the ability of hiding distortions is different in different image regions (Legge and John 1980). Visual masking refers to the effect that the thresholds of the target stimulus increase (the visibility of target reduces) with the occurrence of masking stimulus. In order to find the regions with distinct distortions, the visual masking effects should be taken into consideration. In IQA metrics, the test images are ordinarily considered as the mask while the distortions are supposed to serve as the target to be detected.

Moreover, there are many types of visual masking effects, such as luminance masking and pattern masking. The pattern masking serves as an important marking effect that has a close relation with IQA, as suggested in Chandler (2013). From the

aspect of the anatomy and physiology, luminance masking results from the properties of retinal adaptation (Graham 1989). With the increase of the luminance of
target, the detection thresholds increase as well. Contrast masking, which is a type
of pattern masking, plays an important role in IQA and other image processing
methods (Jones et al. 1995; Zeng et al. 2002). Many experiments have been conducted in researches about contrast masking effects. In general, with the increase of
the contrast of the masker, the threshold of contrast would increase, which means
that the visibility of target reduces. However, there is an exception. In Legge and
John (1980), the authors employ sine-wave grating signals at 2.0 cpd as signal
(target to be detected) and set the masker frequency as 2.8 cpd. Other parameters of
the signal and target are similar. The property of the contrast making is represented
in the threshold versus contrast curves as shown in Fig. 3.7. The result shows that
the threshold of masking detection commonly increases as the contrast of the
masker is increasing. However, under the circumstances of the lower masking
contrast, there is a "dipper effect." This effect works depending on the differences
between the signal and the masker in terms of orientations, spatial frequency, phase,
and other dimensional factors (Fig. 3.7).

3.3.6 Stereoscopic Vision

Stereoscopic vision is another remarkable quality of HVS, as is known to all, natural
scenes receipted by human vision are always stereoscopic, and the stereoscopic
vision is result from the binocular stereopsis mechanism due to the physiological
structure of human eyes. Responses of human eyes are selective in horizonal
direction, so visual scenes are viewed by two eyes from slightly different position
and angle simultaneously and generate two different retinal images; this procedure is
called binocular stereopsis (Garding et al. 1995; Poggio and Poggio 1984), and the
difference between the two retinal images is defined as binocular disparity (Backus
et al. 1999). If the receipted images on the two retinas are similar, then the binocular

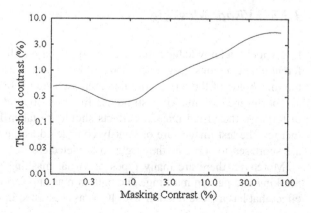

Fig. 3.7 Threshold versus contrast (TvC) curves in Legge and John (1980)

stereopsis can instinctively analyze the difference of two retinal images and construct stereo vision; otherwise, significantly different images will result in binocular rivalry where only one image is perceived (Kaplan and Metlay 1964).

The binocular disparity between the two retinal images is more significant at closer distances. As the distance increases, the disparities get smaller and the ability of human visual system to distinguish differences in depth is declined. In addition, even though human visual system can learn depth information from image receipted by only one eye, stereopsis enhances the ability to discriminate differences in depth and form more comprehensive and more realistic visual effects.

Human vision is well adapted to estimate distance according to the received retinal images. As suggested in Nawrot (2003), the most important depth cues are considered to be motion parallax and binocular stereopsis. The motion parallax cues occur when the observer moves his eyes; it refers to the effect that the fixation point stays on a fixed location and the two images move relatively toward each other on the observer's retina. Binocular depth cues are result from the common situation where two eyes of an observer view the scene at slightly different angles. The mechanism of binocular depth estimation consists of two parts, mergence and stereopsis. The mergence refers to the process in which both eyes take a position to minimize the difference between the retina images, while the stereopsis uses the disparity of the surrounding area for depth estimation.

The concept of stereo vision is the basis of many research fields, such as 3D image processing methods and stereoscopic image quality assessment (SIQA) (Chen et al. 2013; Shao et al. 2013). As a crucial extension of traditional 2D image quality assessment, SIQA is now attracting increasing interests from researchers.

3.4 Summary

With the development in psychology and physiology, people have accumulated an abundant knowledge about human visual system. When we regard the HVS as an information-processing system, with the knowledge of anatomy structures and the properties of HVS, the operating mechanism can be simulated. However, the underlying mechanism of HVS is so complicated that it has not been well studied yet. Meanwhile, still many functions of some cortex cells are undiscovered and many mechanisms are undetermined. Since HVS is a precise system with many levels processing, it is difficult to simulate the HVS completely.

In image quality assessment, the traditional metrics employ the MSE, SNR, and PSNR to measure the errors between original images and distorted images directly. From the 1960s, early researches take the characteristics of human vision into IQA and introduce the visibility error to measure image quality. In Daly model (Daly 1992), the authors tend to calculate the complete differences with CSF filtering, channel decomposition, masking calculation, and other calculation models. The aim of Lubin model (Lubin 1993, 1995) is to find the differences of probability of detection by calculating the various between the different models mentioned before besides PSF filtering. These metrics are intended to have the accurate predictions via building systems to simulate the HVS. However, in Wang and Bovik (2006), the authors put forward four problems toward these models. The four problems refer to the limitations and difficulties of these models from quality definition, suprathreshold problem, natural image complexity, and dependency decoupling. Then, the authors believe that there are many limitations on metrics completely based on HVS.

Although there seems to be no bright future to completely simulate the HVS, the information obtained from HVS is significant for image quality assessment. Many methods in IQA intend to find some sophisticated features that take advantage of the characteristics of HVS to represent the image quality. In consequence, it is convincing that having a great insight into HVS would contribute to make positive progress in IQA.

References

Ahumanda, A. (1996). Simplified vision models for image quality assessment. In *SID International Symposium Digest of Technical Papers*, 97-400.

Alaei, A., Raveaux, R., & Conte, D. (2017). Image quality assessment based on regions of interest. *Signal, Image and Video Processing, 11*(4), 673–680.

Backus, B. T., Banks, M. S., van Ee, R., & Crowell, J. A. (1999). Horizontal and vertical disparity, eye position, and stereoscopic slant perception. *Vision Research, 39*(6), 1143–1170.

Budrikis, Z. L. (1972). Visual fidelity criterion and modeling. *Proceedings of the IEEE, 60*(7), 771–779.

Campbell, F. W., & Robson, J. G. (1968). Application of Fourier analysis to the visibility of gratings. *Journal of Physiology (London), 197*(3), 551–566.

Chandler, D. M. (2013). Seven challenges in image quality assessment: Past, present, and future research. ISRN Signal Processing (pp. 1–53).

Chen, M. J., Su, C. C., Kwon, D. K., Cormack, L. K., & Bovik, A. C. (2013). Full-reference quality assessment of stereopairs accounting for rivalry. *Signal Processing: Image Communication, 28*(9), 1143–1155.

Chen, C., Zhang, X., Wang, Y., Zhou, T., & Fang, F. (2016). Neural activities in V1 create the bottom-up saliency map of natural scenes. *Experimental Brain Research, 234*(6), 1769–1780.

Conway, B. R. (2009). Color vision, cones, and color-coding in the cortex. *The Neuroscientist, 15* (3), 274–290.

Cormack, L. K. (2005). Computational models of early human vision. In *Handbook of image and video processing* (pp. 325–345).

Daly, S. (1992). Visible difference predictor: An algorithm for the assessment of image fidelity. In *Proceedings of SPIE* (Vol. 1616, 2–15).

Daubechies, I., & Sweldens, W. (1998). Factoring wavelet transforms into lifting steps. *Journal of Fourier Analysis and Applications, 4*(3), 245–267.

De Valois, R. L., & De Valois, K. K. (1990). *Spatial vision*. New York: Oxford University Press.

Ding, Y., Zhao, X., Zhang, Z., & Dai, H. (2017). Image quality assessment based on multi-order local features description, modeling and quantification. *IEICE Transactions on Information and Systems, E100-D*(6), 2453–2460.

Felleman, D., & Essen, D. V. (1991). Distributed hierarchical processing in primate cerebral cortex. *Cerebral Cortex, 1*(1), 1–47.

Gao, X., Lu, W., Tao, D., & Li, X. (2009). Image quality assessment based on multiscale geometric analysis. *IEEE Transactions on Image Processing, 18*(7), 1409–1423.

Garding, J., Porrill, J., Mayhew, J., & Frisby, J. (1995). Stereopsis, vertical disparity and relief transformations. *Vision Research, 35*(5), 703–722.

Geisler, W. S., & Banks, M. S. (1995). *Visual performance*. New York: McGraw-Hill Book Company.

Goferman, S., Zelnik-Manor, L., & Tal, A. (2012). Context-aware saliency detection. *IEEE Transactions on Pattern Analysis and Machine Intelligence, 34*(10), 1915–1926.

Gollisch, T., & Meister, M. (2010). Eye smarter than scientists believed: Neural computations in circuits of the retina. *Neuron, 65*(2), 150–164.

Graham, N. (1989). *Visual pattern analyzers*. New York: Oxford University Press.

Gu, K., Zhai, G., Yang, X., & Zhang, W. (2015). Using free energy principle for blind image quality assessment. *IEEE Transactions on Multimedia, 17*(1), 50–63.

Hecht, S. (1924). The visual discrimination of intensity and the Weber-Fechner law. *Journal General Physiology, 7*(2), 235–267.

Itti, L., Koch, C., & Niebur, E. (1998). A model of saliency-based visual attention for rapid scene analysis. *IEEE Transactions on Pattern Analysis and Machine Intelligence, 20*(11), 1254–1259.

Itti, L., & Koch, C. (2000). A saliency-based mechanism for overt and convert shifts of visual attention. *Vision Research, 40*, 1489–1506.

Jones, P. W., Daly, S. J., Gaborski, R. S., & Rabbani, M. (1995). Comparative study of wavelet and discrete cosine transform (DCT) decompositions with equivalent quantization and encoding strategies for medical images. In *Proceedings of SPIE Medical Imaging* (Vol. 2431, pp. 571–582).

Jones, J. P., & Palmer, L. A. (1987). An evaluation of the two-dimensional Gabor filter model of simple receptive fields in cat striate cortex. *Journal of Neurophysiology, 58*(6), 1233–1258.

Kadir, T., & Brady, M. (2001). Saliency, scale and image description. *International Journal of Computer Vision, 45*(2), 83–105.

Kaplan, I. T., & Metlay, W. (1964). Light intensity and binocular rivalry. *Journal of Experimental Psychology, 67*(1), 22–26.

Koch, C., & Poggio, T. (1999). Predicting the visual world: silence is golden. *Nature Neuroscience, 2*(1), 9–10.

Koffka, K. (1955). *Principles of gestalt psychology*. Routledge & Kegan Paul Ltd.

Kottayil, N. K., Cheng, I., Dufaux, F., & Basu, A. (2016). A color intensity invariant low-level feature optimization framework for image quality assessment. *Signal, Image and Video Processing, 10*(6), 1169–1176.

Kruger, N., Janssen, P., Kalkan, S., Lappe, M., Leonardis, A., & Piater, J. (2013). Deep hierarchies in the primate visual cortex: What can we learn for computer vision? *IEEE Transactions on Pattern Analysis and Machine Intelligence, 35*(8), 1847–1871.

Kuffler, S. W. (1953). Discharge patterns and functional organization of mammalian retina. *Journal of Neurophysiology, 16*(1), 37–68.

Legge, G. E., & John, M. F. (1980). Contrast masking in human vision. *Journal of the Optical Society of America, 70*(12), 1458–1471.

Levin, A., & Weiss, Y. (2009). Learning to combine bottom-up and top-down segmentation. *International Journal of Computer Vision, 81*(1), 105–118.

Li, Z. (2002). A saliency map in primary visual cortex. *Trends in Cognitive Sciences, 6*(1), 9–16.

Lin, W., Dong, L. & Xue, P. (2003). Discriminative analysis of pixel difference towards picture quality prediction. In *Proceedings 2003 International Conference on Image Processing* (Cat. No.03CH37429), (Vol. 2, No. 3, pp. 193–196).

Lin, W., & Kuo, C.-C. J. (2011). Perceptual visual quality metrics: A survey. *Journal of Visual Communication and Image Representation, 22*(4), 297–312.

Lubin, J. (1993). The use of psychophysical data and models in the analysis of display system performance. In A. B. Watson (Ed.), *Digital images and human vision* (pp. 163–178). Cambridge: MIT Press.

Lubin, J. (1995). Avisual discrimination mode for image system design and evaluation. *Visual Models for Target Detection and Recognition* (pp. 207–220). Singapore: World Scientific Publishers.

Mannos, J. L., & Sakrison, D. J. (1974). The effects of a visual fidelity criterion on the encoding of images. *IEEE Transactions on Information Theory, 20*(4), 525–536.

Masland, R. H. (2012). The neuronal organization of the retina. *Neuron, 76*(2), 266–280.

Min, X., Zhai, G., Gao, Z., & Gu, K. (2014). Visual attention data for image quality assessment databases. In *2014 IEEE International Symposium on Circuits and Systems (ISCAS)* (pp. 894–897), Melbourne VIC.

Moorthy, A. K., Wang, Z., & Bovik, A. C. (2011). Visual perception and quality assessment. In G. Cristobal, P. Schelkens, & H. Thienpont (Eds.), *Optical and digital image processing*. Weinheim: Wiley Publisher.

Navalpakkam, V., Koch, C., Rangel, A., Perona, P., & Treisman, A. (2010). Optimal reward harvesting in complex perceptual environments. *Proceedings of the National Academy of Sciences of the United States of America, 107*(11), 5232–5237.

Nawrot, M. (2003). Depth from motion parallax scales with eye movement gain. *Journal of Vision, 3*(11), 841–851.

Oliva A. (2005). Gist of the scene. *Neurobiology of Attention,* 251–256.

Orban, G. A. (2008). Higher order visual processing in macaque extrastriate cortex. *Physiological Reviews, 88*(1), 59–89.

Ouria, D. B., Rieux, C., Hut, R. A., & Cooper, H. M. (2006). Immunohistochemical evidence of a melanopsin cone in human retina. *Investigative Ophthalmology & Visual Science, 47*(4), 1636–1641.

Poggio, G., & Poggio, T. (1984). The analysis of stereopsis. *Annual Review of Neuroscience, 7*(1), 379–412.

Saha, A., & Wu, Q. M. J. (2016). Full-reference image quality assessment by combining global and local distortion measures. *Signal Processing, 128,* 186–197.

Sakrison, D., & Algazi, V. (1971). Comparison of line-by-line and two-dimensional encoding of random images. *IEEE Transactions on Information Theory, 17*(4), 386–398.

Schade, O. H. (1956). Optical and photoelectric analog of the eye. *Journal of the Optical Society of America, 46*(9), 721–739.

Schreiber, W. F. (1986). *Fundamentals of electronic imaging systems*. Berlin: Springer.

Shao, F., Lin, W., Gu, S., Jiang, G., & Srikanthan, T. (2013). Perceptual full-reference quality assessment of stereoscopic images by considering binocular visual characteristics. *IEEE Transactions on Image Processing, 22*(5), 1940–1953.

Shapley, R., & Hawken, M. J. (2011). Color in the cortex: Single- and double-opponent cells. *Vision Research, 51*(7), 701–717.

Shen, D., & Wang, S. (1996). Measurements of JND property of HVS and its applications to image segmentation, coding and requantization. In *Proceedings of SPIE* (Vol. 2952, pp. 113–121).

Stockham, T. G. (1972). Image processing in the context of a visual model. *Proceedings of the IEEE, 60*(7), 828–842.

Tatler, B. W., Wade, N. J., Kwan, H., Findlay, J. M., & Velichkovsky, B. M. (2010). Yarbus, eye movements, and vision. *I-Perception, 1*(1), 7–27.

Taylor, C., Pizlo, Z., Allebach, J. P., & Bouman, C. A. (1997). Image quality assessment with a Gabor pyramid model of the human visual system. In *Proceeding of SPIE* (Vol. 3016, pp. 58–69).

Tong, Y. B., Konik, H., Cheikh, F. A., & Tremeau, A. (2010). Full reference image quality assessment based on saliency map analysis. *Journal of Imaging Science and Technology, 54* (3), 305031–305034.

Treisman, A., & Gelade, G. (1980). A feature-integration theory of attention. *Cognitive Psychology, 12*(1), 97–136.

Vu, C. T., Larson, E. C., & Chandler, D. M. (2008). Visual fixation patterns when judging image quality: Effects of distortion type, amount, and subject experience. In *Proceedings of the IEEE Southwest Symposium on Image Analysis and Interpretation (SSIAI '08)* (pp. 73–76).

Wandell, B. A. (1995). *Foundations of vision*. Sinauer Associates, Inc.

Wang, Z., & Bovik, A. C. (2006). Modern image quality assessment. *Synthesis Lectures on Image, Video, and Multimedia Processing, 2*(1), 1–156.

Watson, A. B. (1993). DC Tune: A technique for visual optimization of DCT quantization matrices for individual images. In *Society for Information Display Digest of Technical Papers* (Vol. XXIV, 946–949).

Watson, A. B., & Ahumanda, A. (2005). A standard model for foveal detection of spatial contrast. *Journal of Vision, 5*(9), 717–740.

Watson, A. B., Hu, J., & McGowan, J. F., III. (2001). DVQ: A digital video quality metric based on human vision. *Journal of Electronic Imaging, 10*(1), 20–29.

Watson, A. B., Yang, G. Y., Solomon, J. A., & Villasenor, J. (1997). Visibility of wavelet quantization noise. *IEEE Transactions on Image Processing, 6*(8), 1164–1175.

Wilson, H. R., & Regan, D. (1984). Spatial frequency adaptation and grating discrimination: Predictions of a line element model. *Journal of the Optical Society of America A: Optics and Image Science, and Vision, 1*(11), 1091–1096.

Winkler, S. (1999). A perceptual distortion metric for digital color video. In *Proceedings of SPIE* (Vol. 3644, 175–184).

Wolfe, J. (1994). Guided search 2.0: A revised model of visual search. *Psychonomic Bulletin & Review, 1*(2), 202–238.

Wu, J., Lin, W., Shi, G., & Liu, A. (2013). Perceptual quality metric with internal generative mechanism. *IEEE Transactions on Image Processing, 22*(1), 43–54.

Wu, H. R., & Rao, K. R. (2006). *Digital video image quality and perceptual coding*. Taylor & Francis.

Xue, W., Zhang, L., Mou, X., & Bovik, A. C. (2014). Gradient magnitude similarity deviation: A highly efficient perceptual image quality index. *IEEE Transactions on Image Processing, 23*(2), 684–695.

Yamada, K., & Cottrell, G. W. (1995). A model of scan paths applied to face recognition. In *Proceedings of the 17th Annual Conference of the Cognitive Science Society* (pp. 55–60).

Zeng, W., Daly, S., & Lei, S. (2002). An overview of the visual optimization tools in JPEG 2000. *Signal Processing: Image Communication, 17*(1), 85–104.

Zhang, L., Shen, Y., & Li, H. (2014). VSI: A visual saliency-induced index for perceptual image quality assessment. *IEEE Transactions on Image Processing, 23*(10), 4270–4281.

Zhang, L., Tong, M. H., Marks, T. K., Shan, H., & Cottrell, G. W. (2008). SUN: A bayesian framework for saliency using natural statistics. *Journal of Vision, 8*(7), 32.

Chapter 4
General Framework of Image Quality Assessment

4.1 Introduction

In this chapter, we are going to provide a rough introduction about the general framework of modern image quality assessment (IQA) methods. A basic knowledge about IQA that should be aware in advance is that the most common classifying scheme for IQA is by three categories, *full-reference* (FR), *reduced-reference* (RR), and *no-reference* (NR). As the terms imply, this classification is according to whether "reference" is involved during IQA operation (Wang and Bovik 2006). When we are assessing the quality of an image, we define its *reference image* (or simply its *reference*) as its undistorted version. The reference image is also called *original image* in some literature. That means, for an assessed image, we always regard that it is somewhat distorted from its reference; if the distortion degree is zero, the image is considered with perfect quality.

For IQA methods belonging to different categories among the three, the framework they follow might slightly different. We will spend the rest of this section introducing each of them. On the other hand, of course, the basic principle of them is the same. Concretely, no matter which type of methods we are designing, it is always required to extract quality-aware features from images and then quantify the features and map the results to the final image quality scores. These common procedures will be discussed in detail in Sects. 4.2 and 4.3, and some of more specific contents are even contained in Chaps. 5 and 6. Also evidently, the statistical metrics that measure the performance of IQA methods are universal for them, and the corresponding contents will be introduced in Sect. 4.4.

© Zhejiang University Press, Hangzhou and Springer-Verlag GmbH Germany 2018   45
Y. Ding, *Visual Quality Assessment for Natural and Medical Image*,
https://doi.org/10.1007/978-3-662-56497-4_4

4.1.1   Typical Framework of Full-Reference Methods

When image quality is assessed by an objective IQA method, if the complete knowledge about the corresponding reference is required, the method is referred to as FR. Then, each tested image contains exactly same content as its reference, only with potential quality degradation caused by distortions. The degree of quality degradation can be reflected by the difference between the tested and reference images. The relationship between them is illustrated in Fig. 4.1, which is also the basis of FR IQA theory. It is not practical to directly measure the degree of distortion. On one hand, it is hard to successfully do so, and we are even not clear what distortion type the image is suffering from in most cases. On the other hand, it is not easy to separate the distortion from image content except for some special distortions such as additive noise. Luckily, for FR methods, we can measure the degree of distortion or image quality degradation by the difference between the tested and reference images. It is worth noted that the relationship 1 in Fig. 4.1, i.e., distortion causes image quality degradation and the latter reflects the former, always hold; while the relationship 2 is useful only for FR methods, because for RR and NR methods, information about the references is not entirely available.

So now we transfer IQA to a more concrete problem, to quantify the difference between the tested and reference images. There are numerous ways to find the difference between two signals, but the desirable ways for IQA must be consistent with human's subjective perception upon image quality. We hope to derive a better objective quality score for an image that has perceptually smaller disparity to its reference; and how to measure the perceptual disparity is not arbitrary. For instance, in the very early stage of IQA, mean squared error (MSE) is one popular method, which is computed by

$$\text{MSE} = \frac{1}{N} \sum_{i=1}^{N} \left(\mathbf{T}(i) - \mathbf{R}(i) \right)^2 \tag{4.1}$$

where $\mathbf{T}$ and $\mathbf{R}$ denote the grayscale images of the tested and reference images, and N is the total pixel number in the tested image. Since all pixels in tested and reference images are used for computation, MSE is undoubtedly an FR method. The computing process is rather simple, the Euclidean distance (which will be introduced in detail in Sect. 4.3) of the grayscale matrices, divided by the data scale, which is regarded as the quality indicator. MSE, as a distance measuring tool, is certainly capable to measure the disparity between the tested and reference

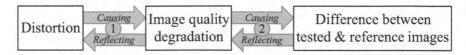

Fig. 4.1 Theoretical basis for FR methods: relationships between image distortion, image quality degradation, and the difference between the tested and reference images

images, but the disparity is not our perceptual quality difference for images, which is why the accuracy of MSE is widely criticized (Wang and Bovik 2009).

In order to find proper ways to quantify the abovementioned perceptual disparity, we have to describe images in some form that better reflects their quality, instead of a pixel-based manner. A more former and popular expression is to extract *quality-aware* features from images (Wang et al. 2006). Quality-aware is a property denoting that the features correlate with perceptual image quality very well. In recent years, many image features are quality-aware such as luminance, gradient, information content, local energy, local texture (Sheikh and Bovik 2006; Zhang et al. 2011; Liu et al. 2012; Ding et al. 2014; Zhang et al. 2014; Chang et al. 2015). Later, the disparity measurement is operated on the features rather than raw images. Since it is currently acknowledged that a comprehensive image quality description is composed of many different image feature representation, we are very likely to obtain multiple disparity measuring results. Therefore, the final step is to map these results into one objective quality score for the tested image. In summary, we roughly divide typical FR IQA operating procedure into three serial stages, of which the framework is given in Fig. 4.2.

4.1.2 Typical Framework of Reduced-Reference Methods

RR is perhaps the most confusing category among the three. It can be regarded as a trade-off between FR and NR. The merits and flaws for FR and NR methods are both obvious. FR methods make the best use of the reference images and generally achieve comparatively best performance according to the literature. However, the scenarios that reference images are fully available are rather rare. On the contrary, NR methods are applicable in almost all situations in reality, but the predicting accuracy is relatively low. The introduction of RR methods creates a new approach for IQA. For certain application scenarios, prior information about the reference images is partly known. If the known information has potential for reflecting image quality, we can extract quality-aware features from it and apply RR IQA methods to evaluate the image quality, which can take advantage of more prior knowledge than NR methods and therefore achieve better results, and do not have too strict requirements as FR ways (Li and Wang 2009).

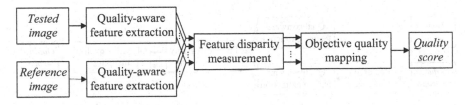

Fig. 4.2 Typical framework of FR methods

RR methods are highly scenario-dependent. In different application scenarios, the available reference information might vary, so different methods should be applied. For a specific RR method, we can consider the process as if the reference images go through a channel that operates similar to a filter. Then, the quality-aware features are only possibly extracted from the retained reference image content. A typical framework of RR methods is shown in Fig. 4.3.

Comparing to the flow shown in Fig. 4.2, the only explicit modification in Fig. 4.3 is the additive process to restore the reference images. Because we have to design different RR methods for different applications, RR methods are classified into many different types accordingly, and the classification itself is quite complicated. How to select proper situations that let RR methods be maximally useful is a trick problem, and in the literature, study upon RR methods is much fewer than FR and NR methods.

4.1.3 Typical Framework of No-Reference Methods

The concept of NR (another popular term is *blind*, denoting exactly the same thing) is very easy to understand. When we are employing an NR method, we require no prior knowledge other than the tested image itself. Thus, NR solutions seem to be the ultimate goal of IQA. Without the reference images, it is not possible for us to find the "disparity" between the quality-aware features of the tested and reference images. But capturing quality-aware features is still an inevitable step. Except for feature quantification, other procedures of NR methods are very similar to those of FR. So, the essential problem is how to build the connection between quality-aware features and quality indices in the absence of references.

There are generally two kinds of approaches that are commonly adopted in existing works. The first kind is to build artificial reference images. This kind of methods can also be divided into two subcategories. Some methods construct one ideal model and regard the difference between a test image and this ideal model as its quality (Moorthy and Bovik 2011; Saad et al. 2012), of which the typical framework is shown in Fig. 4.4. These methods assume that the undistorted natural images share similar statistics (Simoncelli and Olshausen 2001). The others adopt denoising schemes to try to recover the corresponding reference for each tested image (Portilla et al. 2003). Because the denoising algorithms for different

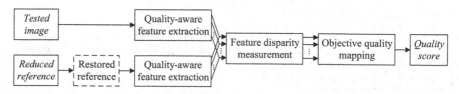

Fig. 4.3 Typical framework of RR methods

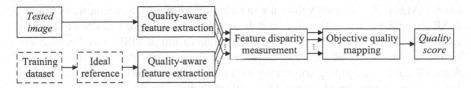

Fig. 4.4 Typical framework for NR methods by constructing an "ideal" reference

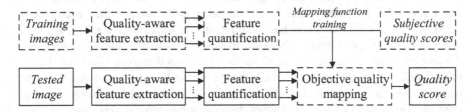

Fig. 4.5 Typical framework of NR methods by training quality mapping function

distortion types can be very different, often a classifier is set beforehand to predict the distortion types (Moorthy and Bovik 2010). In fact, in the early stages of NR study, there have been proposed lots of methods that are distortion-specific (Wu and Yuen 1997; Wang et al. 2000; Yu et al. 2002; Wang et al. 2002; Sheikh et al. 2002; Marziliano et al. 2002; Sheikh et al. 2005). The NR methods for general purposes are only largely emerging after around 2010s.

The second kind of methods appears more recently, which directly map the quality-aware features to the final quality scores, without attempting to build pseudo-references, as shown in Fig. 4.5. Obviously, besides the coherency between the features and human subjective perception, the performance of such methods is also largely dependent on the quality of the mapping functions and the training databases. But, since the disparity measurement is omitted, these methods do not rely on proper distance quantification schemes and are usually computationally faster. Therefore, the second kind is becoming more and more popular in recent years. However, we have to emphasize that it is not ensured that which of the two is better; it is really depending on specific methods; with these difficulties involved, NR methods are generally exhibiting comparatively worst performance than FR methods, even in the current stage.

4.2 Quality-Aware Feature Extraction

Extracting quality-aware features from images is a necessary step for any FR, RR, and NR methods, although sometimes not an explicit one. For the simplest case, again taking MSE as an example, the features are the grayscale values of image pixels. Although grayscale values are not considered quality-aware now, as long as MSE is utilized for IQA in a certain application, they are acknowledged as quality-aware by the utilizers. Also, it is unfair to judge MSE now because our definition of quality-aware may be changing over time. Perhaps, many image features regarded as highly correlating with quality will not be considered so in near future. For a more complex example, methods built upon the convolutional neural network (CNN) can learn to capture effective features by themselves. If the inputs for CNN are directly set as image pixels, the feature extraction, feature quantification, and quality mapping are integrated as one procedure, but it is safe to state that each of the three procedures is still important and cannot replace each other.

Moreover, feature extraction has more significant importance for IQA than the rest parts. It tells the difference between methods. Designing new IQA method is the process to find a feature or a combination of features that more effectively reflect image quality than existing ones. In comparison, the left stages are just trying to make the best of the features to build as strong connection to final quality scores as possible. On the other hand, the feature disparity measurement, quantification, and mapping schemes are mostly borrowed from other related fields such as statistics and machine learning, and the useful approaches are quite finite, yet the feature extraction schemes, from today's prospective, can be rather extensive, especially with extension through combinations.

Because of the number of feature types are so huge, it is not practical to introduce all of them, and it is even not possible to offer a strict classification to divide them. In this book, we roughly define two categories: the methods based on human visual system (HVS) and those based on natural scene statistics (NSS), which will be discussed detailed in Chaps. 5 and 6, respectively. We are merely giving a very brief introduction in the following paragraphs.

HVS-based methods are developed directly based on the knowledge about how human perceives quality. HVS is a highly complex hierarchical nonlinear system. Each hierarchy in HVS is approximately a band-pass filter that receives, integrates, and processes visual signals from its former layer and outputs the processing results to the next one (Wandell 1995; Geisler and Banks 1995; Cormack 2005). Some early researchers of modern IQA have literally tried to model the transforming characteristics of each hierarchy in HVS by constructing complex mathematical functions (Safranek and Johnston 1989; Rao and Yip 1990; Daly 1992; Lubin 1993; Teo and Heeger 1994). We now know that our current understanding and hardware functionality are both far from adequate to build such models. To better simulate HVS functionalities, sophisticated models have to be developed. So, it is highly restrained by our knowledge level in neurosciences, which is still quite immature nowadays. In addition, the simulation can be very computationally costly and inconvenient. Therefore, more recent HVS-based methods only simulate one or a few quality-aware properties of HVS so that computational complexity is largely saved and the whole process becomes much more concise controllable. NSS methods overcome the obstacles we meet in studying HVS models by cutting a short path. It is assumed that images with perfect quality always obey certain statistical laws. Another way to put it, the natural scenes have to be statistically regular so that they can be appreciated by HVS. Therefore, directly studying how the regularity is like gives us another thought for IQA but avoids the complex simulation for HVS.

However, we have to emphasize that classification strategy of IQA methods according to whether the feature extraction schemes are based on HVS or NSS is far from perfectly accurate. We are doing so in the book mainly because it helps us to introduce existing methods more orderly. There are no explicit boundaries between the two categories: HVS-based methods try to explain what kind of images makes our visual cortex most comfortable, and the scenes from these images must coherent the best with the statistical regularities and vice versa; they are just two different starting points to search the solutions for the same problem and will find peace with each other if we are really approaching the ground-truth answers.

4.3 Feature Quantification and Mapping Strategy

4.3.1 Feature Quantification

The extracted features have to be quantified so that they can reflect the objective image quality. For simplicity, we will mainly discuss feature extraction from grayscale images. Color images occupy two more channels than gray one, so if an adopted feature extraction tool is also sensitive to color, the obtained features will carry three times information comparing to the gray images; but in reality, lots of feature extraction tools are not color-aware, so applying them for color images will result in similar effects as gray images. Thus, mere introduction about grayscale image feature extraction will not lose generality.

We can basically divide the features into three types: pixel-wise, patch-wise, and global. For a grayscale image, pixel-wise feature extraction tools will generate a feature matrix of similar size, with each item corresponding to each pixel. While patch-wise tools develop features conveying less amount of information, they still have to be represented as a matrix or vector containing items far more than one. Comparatively, global features represent one aspect of an image with only one or a few indices.

From the aspect of quantization, we are more pleased to deal with less amount of data. For instance, assuming we use one number to denote the luminance level of an image and one number for its corresponding reference, then for NR methods, we can directly use this number as a quality index representing image luminance; for FR methods, we can directly find the difference between the two numbers to act as a quality index. However, from the aspect of feature extraction, pixel-wise and patch-wise features would be preferred, because they provide a more detailed reflection of what we capture. Therefore, they are much more commonly employed in practice.

Assuming the image we are processing is of size $M \times N$, and k numbers are utilized to represent the quality-aware features extracted by a pixel-wise or patch-wise tool. Usually, k will satisfy

$$0 < k \leq M \times N \tag{4.2}$$

We use a vector with length k to envelop the data. For NR methods, it is theoretically correct to train a regression function that maps the k indices of a tested image, enveloped in vector $\mathbf{t}$, to its quality score using machine learning techniques. But since k may be rather huge, this approach is not really feasible. In practice, usually it is some statistics of $\mathbf{t}$, e.g., its mean, median, variation, that are computed and mapped to the quality score. For FR or RR method, similarly, the difference between features from the tested image and its reference is preferably represented by some approaches that use very limited number to quantify the similarity or dissimilarity between them, rather than the assemble of differences between each corresponding item. Name the feature vector of a reference image as $\mathbf{r}$; how to properly measure the difference between $\mathbf{t}$ and $\mathbf{r}$ is very important. Since there exist lots of measuring approaches for us to choose, we will give a brief overview of most common one. We have to point out that the following discussion aims to illustrate only the basic concept of feature distance quantification, instead of a comprehensive overview. This discussion is well within the scope of statistics, so to build more solid understanding, readers are recommended to pick up some related materials, such as Mood et al. (1974), Casella and Berger (2001), Larsen and Marx (2005).

The features are represented by numbers or nominal values depending on specific situation, where the former is seen in most cases. So, we are starting with them. L_p distances are very widely utilized, defined as

$$L_p(\mathbf{t}, \mathbf{r}) = \left(\sum_{i=1}^{k} |t_i - r_i|^p \right)^{1/p} \tag{4.3}$$

where t_i and r_i denote the items in $\mathbf{t}$ and $\mathbf{r}$, respectively. Parameter p is usually set as 1, 2, or infinite. When p is set, the specific L_p distance is referred to as L_1 distance, L_2 distance, or L_∞ distance,

$$L_1(\mathbf{t}, \mathbf{r}) = \sum_{i=1}^{k} |t_i - r_i| \tag{4.4}$$

$$L_2(\mathbf{t}, \mathbf{r}) = \sqrt{\sum_{i=1}^{k} (t_i - r_i)^2} \tag{4.5}$$

$$L_\infty(\mathbf{t}, \mathbf{r}) = \max|t_i - r_i| \tag{4.6}$$

L_p distance, L_1 distance, L_2 distance, and L_∞ distance, are also referred to as Minkowski distance, Manhattan distance, Euclidean distance, and Chebyshev distance, respectively.

Cosine similarity (CS) is also widely used. It measures the intersection angle of the two lines connecting the original point, and the points **t** and **r** represent in the k-dimensional space,

$$CS(\mathbf{t}, \mathbf{r}) = \frac{\sum_{i=1}^{k} t_i r_i}{\sqrt{\sum_{i=1}^{k} t_i \sum_{i=1}^{k} r_i}} \tag{4.7}$$

Because it quantifies the angle but ignores the ground-truth distance, cosine similarity is often combined with Euclidean distance.

The distance of two vectors composed by nominal values is usually computed by finding the difference between the two statistical distributions. The distributions are often in the form of histograms. So, we are actually measuring the difference between two histograms. Assuming there are b possible candidates of the nominal values, **t** and **r** can be represented as histograms **ht** and **hr**, both with b items; the value of each item in a histogram is an integer denoting how many times it occurs in the original vector. Sometimes the normalized version of histograms is adopted, in which each item is the original value divided by k. The normalized histograms are sometimes more convenient for computation, but the two versions share the same essence and results in most cases. Common solutions for computing the distance between two histograms include chi-square (χ^2) distance, histogram intersection (HI) distance, Bhattacharyya (BH) distance,

$$\chi^2(\mathbf{ht}, \mathbf{hr}) = \sum_{i=1}^{b} \frac{(ht_i - hr_i)^2}{ht_i + hr_i} \tag{4.8}$$

$$HI(\mathbf{ht}, \mathbf{hr}) = \sum_{i=1}^{b} \min(ht_i, hr_i) \tag{4.9}$$

$$BH(\mathbf{ht}, \mathbf{hr}) = \sqrt{1 - \sum_{i=1}^{b} \frac{\sqrt{ht_i, hr_i}}{\sum_{i=1}^{b} ht_i \sum_{i=1}^{b} hr_i}} \tag{4.10}$$

where ht_i and hr_i denote the value of bin i in **ht** and **hr**, respectively. A suggestion is that when accuracy is very much pursued, adopt chi-square distance or Bhattacharyya distance; while if efficiency is preferred, use histogram intersection.

The above gives some very simple illustration of distance measurement for vectors. In practice, there are extensive choices of such quantifying approaches, created for matrices, vectors, or numbers. To give conclusions about how to decide which one to use is difficult because the problem is highly depending on situations. Therefore, the quantification of features is basically an experimental-oriented issue.

4.3.2   Mapping Strategy

The majority of modern IQA methods employ more than one quality-aware feature to depict image quality. So, multiple indices will be generated, and the final quality scores are the synthesis of them all. Supposing there are n indices $\{x_1, x_2, \ldots, x_n\}$, the final quality score q is computed with a regression function f,

$$q = f(x_1, x_2, \ldots, x_n) = f(\mathbf{x}) \tag{4.11}$$

where $\mathbf{x}$ is the index vector containing the n indices. Using the form of vectors and matrices has exactly the same effects as numbers but will allow us to illustrate more clearly and help understanding with linear algebra.

How to find f is a problem belonging to the field of machine learning, with plenty of possible solutions, e.g., linear regression (LR), k-nearest neighbor (kNN), support vector regression (SVR), neural networks (NN). We will briefly introduce the listed four approaches, because they are the mostly adopted tools for regression problems, not only within IQA, but also for many other applications. The output q for training denotes the subjective scores of images. With the derived f, objective scores q' can be developed.

Before starting any of these approaches, feature normalization is the first thing that should be taken care of. Because the features are captured using various means, the indices are possibly in very different scales. Data preprocessing for machine learning usually normalizes the values to the range of $[-1, 1]$ or $[0, 1]$. The specific normalization strategy varies but the results are very similar, so we are not digging into it. Say we have m images to train f, then each index vector $\mathbf{v}$ is of length m. A simple way to normalize $\mathbf{v}$ to $[0, 1]$ is

$$v_i = \frac{v_i - \min(\mathbf{v})}{\max(\mathbf{v}) - \min(\mathbf{v})} \tag{4.12}$$

where v_i are the items of $\mathbf{v}$.

LR and kNN are the simplest regression model. If LR is applied, q is regarded as the weighted sum of the indices, with an additional bias b,

$$q = \sum_{i=1}^{n} w_i x_i + b \tag{4.13}$$

Then our objective is to find the proper w_i and b. Rewriting it using the form of linear algebra will help understand it. Let $\mathbf{q}$ be a vector with length m containing quality scores of the m images, $\mathbf{X}$ be a matrix of size $m \times (n + 1)$, containing the indices of every images with an additional column to correspond to b, and $\mathbf{w}$ be a $(n + 1)$ vector including w_i and b, i.e.,

$$\mathbf{q} = \begin{bmatrix} q_1 \\ q_2 \\ \vdots \\ q_m \end{bmatrix}, \quad \mathbf{X} = \begin{bmatrix} \mathbf{x}_1^T & 1 \\ \mathbf{x}_2^T & 1 \\ \vdots \\ \mathbf{x}_m^T & 1 \end{bmatrix}, \quad \mathbf{w} = \begin{bmatrix} w_1 \\ w_2 \\ \vdots \\ w_n \\ b \end{bmatrix} \tag{4.14}$$

Then (4.13) can be rewritten as

$$\mathbf{q} = \mathbf{X}\mathbf{w} \tag{4.15}$$

The above equation does not always hold except for an ideal linearly related $\mathbf{X}$ and $\mathbf{q}$, which is not possible in reality. So, our goal is to find the best w, w* that minimizes the nonlinearity of the relation between them, i.e.,

$$\mathbf{w}^* = \arg\min_{\mathbf{w}} (\mathbf{q} - \mathbf{X}\mathbf{w})^T (\mathbf{q} - \mathbf{X}\mathbf{w}) \tag{4.16}$$

which can be analytically resolved by

$$\mathbf{w}^* = (\mathbf{X}^T \mathbf{X})^{-1} \mathbf{X}^T \mathbf{y} \tag{4.17}$$

The inverse matrix of $\mathbf{X}^T\mathbf{X}$ can be found by the Moore–Penrose algorithm and pseudo-inverse computation algorithm (Moore 1920; Bjerhammar 1951; Penrose 1955) when $\mathbf{X}^T\mathbf{X}$ is not a full-rank matrix.

The theory of kNN is even simpler. The quality score of a tested sample is regarded as the average of its k-nearest neighboring training samples (Xu et al. 1992; Ho et al. 1994; Breiman 1996; Kittler et al. 1998). In practice, kNN and LR perform worse than SVR and NN, because their simplicity decides that they are not capable to deal with complex situations. However, they have the advantage of avoiding the "learning" process that is very costly in time for SVR and NN, especially the latter. Therefore, in scenarios that time is very precious but inaccuracy can be tolerated to some degree, kNN and LR are recommended. Generally speaking, LR tends to achieve relatively better results, and also because there is no hyperparameter that needs to be set beforehand for LR (while k needs to be determined for kNN), LR is more often employed than kNN.

In addition, LR can be extended to other forms. The exponential regression is a commonly seen example, which can be expressed as

$$q = \prod_{i=1}^{n} x_i^{w_i} \tag{4.18}$$

Using its logarithm

$$\log q = \sum_{i=1}^{n} w_i \log x_i \qquad (4.19)$$

easily transfers it to an LR problem.

SVR is the regression function of support vector machine (SVM) (Cortes and Vapnik 1995). The earliest SVM is developed for binary classification; the classification function of SVM is also referred to as support vector classification (SVC). The binary SVC aims to build a hyperplane in the n-dimensional space (n is the number of indices as above),

$$\mathbf{w}^{\mathrm{T}}\mathbf{x} + b = 0 \qquad (4.20)$$

that best distinguishes the two classes, which is desired to be as far from the samples on its either side as possible and classify the points correctly, equaling to the optimization problem

$$\min_{\mathbf{w},b} \frac{1}{2}\|\mathbf{w}\|^2 + C \sum_{i=1}^{m} \ell_{0/1}(\mathbf{x}_i, q_i) \qquad (4.21)$$

where $\|\mathbf{w}\|^{-2}$ is the distance between the hyperplane and the support vectors (the points that are nearest to the hyperplane), $\ell_{0/1}$ is the loss function for binary classification that is positive value when sample i is correctly classified and 0 otherwise, and C is a hyperparameter that balances the two separate optimization goals. For regression problem, we only have to replace $\ell_{0/1}$ by a loss function ℓ_ε to measure the inaccuracy of the predicted quality scores, so the optimization object is

$$\min_{\mathbf{w},b} \frac{1}{2}\|\mathbf{w}\|^2 + C \sum_{i=1}^{m} \ell_\varepsilon(\mathbf{x}_i, q_i) \qquad (4.22)$$

The most widely adopted loss function is ε-insensitive loss,

$$\ell_\varepsilon(\mathbf{x}, q) = \max(0, |f(\mathbf{x}) - q| - \varepsilon) \qquad (4.23)$$

which means when the predicted quality score $f(\mathbf{x})$ differs from q within ε, the loss is 0. Introducing slack variables ξ, the optimization object can be rewritten as

$$\min_{\mathbf{w},b} \frac{1}{2}\|\mathbf{w}\|^2 + C \sum_{i=1}^{m} \ell_\varepsilon(\mathbf{x}_i, q_i)$$

$$\text{s.t.} \begin{cases} f(\mathbf{x}_i) - y_i \leq \varepsilon + \xi_i, \\ y_i - f(\mathbf{x}_i) \leq \varepsilon + \widehat{\xi}_i, \\ \xi_i \geq 0, \widehat{\xi}_i \geq 0 \end{cases} \qquad (4.24)$$

which can be solved by the method of Lagrange multipliers. By introducing the Lagrange multipliers, the problem is turned into its dual problem,

$$\max_{\widehat{\alpha}, \alpha} \sum_{i=1}^{m} q_i(\widehat{\alpha}_i - \alpha_i) - \varepsilon(\widehat{\alpha}_i + \alpha_i) - \frac{1}{2}\sum_{i=1}^{m}\sum_{j=1}^{m}(\widehat{\alpha}_i - \alpha_i)(\widehat{\alpha}_j - \alpha_j)\mathbf{x}_i^{T}\mathbf{x}_j$$

$$\text{s.t.}\begin{cases} \sum_{i=1}^{m}(\widehat{\alpha}_i - \alpha_i) = 0, 0 \leq \alpha_i, \widehat{\alpha}_i \leq C, \alpha_i\widehat{\alpha}_i = 0, \xi_i\widehat{\xi}_i = 0 \\ \alpha_i(f(\mathbf{x}_i) - q_i - \varepsilon - \xi_i) = 0, (C - \alpha_i)\xi_i = 0, \\ \widehat{\alpha}_i\left(q_i - f(\mathbf{x}_i) - \varepsilon - \widehat{\xi}_i\right) = 0, (C - \widehat{\alpha}_i)\widehat{\xi}_i = 0 \end{cases} \quad (4.25)$$

which is a quadratic programming program that can be solved by many ways, of which sequential minimal optimization (Platt 1998) is a suggested one.

However, the above analysis assumes that the samples in the n-dimensional hyperspace are linear separable. But the assumption fails for almost all real-world scenarios. So, the indices $\mathbf{x}_i$ are always transferred to higher-dimensional spaces using a projecting function φ, and the optimization involves finding the results of $\varphi(\mathbf{x}_i)^{T}\varphi(\mathbf{x}_j)$. Because it will be very computationally complex to find the inner product of vectors with such high dimension, kernel trick is invented. A kernel function is defined as

$$\kappa(\mathbf{x}_i, \mathbf{x}_j) = \varphi(\mathbf{x}_i)^{T}\varphi(\mathbf{x}_j) \quad (4.26)$$

Linear kernel, polynomial kernel, Gaussian kernel, Laplacian kernel, and sigmoid kernel are several common kernel functions for SVM. The most widely adopted one is Gaussian kernel, also referred to as radius basis function (RBF) kernel,

$$\kappa(\mathbf{x}_i, \mathbf{x}_j) = \exp\left(-\frac{\|\mathbf{x}_i - \mathbf{x}_j\|^2}{2\sigma^2}\right) \quad (4.27)$$

Employing kernel functions is a standard procedure for modern SVM and NN.

NN tries to simulate human brain by employing a hierarchical structure, which is composed by hierarchically ordered computation units called "neurons" and their connections. Each hierarchy is called a "layer". One NN contains an input layer, an output layer, and any number of hidden layers between the input and output layers. Signals are inputted from the input layer and transmitted to the output layer through each hidden layer. Except for those in the input layer, all neurons in a layer receive signals from neurons in its previous layer and send signal to neurons in its next layer. For a neuron, supposing there are p neurons in its previous layer outputting signals s_i ($i = 1$ to p), then the signal of this neuron outputs, o, would be

$$o = g\left(\sum_{i=1}^{p} w_i s_i + b\right) \tag{4.28}$$

where w_i and b are parameters to be learned, and g is an activation function that "activate" the network so that it becomes nonlinear and the performance will be largely improved. Most widely adopted activation functions are sigmoid function, tanh function, rectified linear unit, etc. The training of NN is the process to find the best fit parameters w_i and b using the famous back propagation (BP) algorithm. Readers can refer to the literature in machine learning field for more information (Haykin 1998); we are going into it too deep.

SVM and NN are very much correlated with each other. If the number of neurons in the only hidden layer of a certain NN is set as the number of training samples, then the NN with RBF as activation function is exactly the same as an SVM with RBF kernel in functionality.

Theoretically, an NN with quantitatively sufficient neurons in its hidden layers is capable to learn all possible mapping relations, even if the relations are extremely complex and only one hidden layer is embedded, but adding hidden layers is obviously much more efficient than increasing neuron numbers. Usually, we call a learning process with NNs containing more than three layers as "deep learning".

In practice, too large quantity of hidden layers arouses some problems. For one thing, the computational complexity is increasing very fast corresponding, leading to huge time consuming. There are many approaches to overcome the problem, though. For two well-known examples, unsupervised layer-wise training and weight sharing are good thoughts and successfully applied in deep belief network (DBN) (Hinton et al. 2006) and convolutional neural network (CNN) (LeCun and Bengio 1995; LeCun et al. 1998), which have both been widely appraised and employed. For another, too complex networks will bring about over-fitting, which are very commonly occurred in machine learning, especially when the samples are not sufficient and the network structure is too complex. For an IQA method, suggestively, when the samples are limited in number, SVR can be both accurate and efficient; while if extensive samples are available, NN with more layers can achieve better performance.

4.4 Statistical Evaluation of Objective Methods

The construction of IQA databases provides a reliable way to evaluate objective IQA methods fairly by regarding the correlation degree of subjective scores and objective scores of a set of images as how good the objective method is. We have plenty of statistical tools to measure the correlation between two sets of data.

Linear correlation is the simplest way to compute the degree of correlation. However, linear correlation computation requires the input data to be perfectly linear correlated. For objective IQA methods, it is hard to ensure that the data contribution follows similar regularity with subjective scores. For the sake of fairness, builders of LIVE IQA databases (Sheikh et al. 2006) suggest employing a nonlinear mapping for the objective scores before computing linear correlation,

$$q' = \beta_1 \left(\frac{1}{2} - \frac{1}{1 + \exp(\beta_2(q - \beta_3))} \right) + \beta_4 q + \beta_5 \qquad (4.29)$$

where q and q' are the objective scores before and after the mapping. In most cases, q' can be a fair index for linear correlation computation.

Pearson's linear correlation coefficient (PLCC), also denoted as correlation coefficient (CC), is a basic metric for measuring linear correlation. Let $\mathbf{x}$ and $\mathbf{y}$ denote the vectors containing objective and subjective quality scores of m images, and x_i and y_i denote their contained items, respectively, then PLCC can be expressed as

$$PLCC(\mathbf{x}, \mathbf{y}) = \frac{\sum_{i=1}^{m} (x_i - \bar{x})(y_i - \bar{y})}{\sqrt{\sum_{i=1}^{m} (x_i - \bar{x})^2} \sqrt{\sum_{i=1}^{m} (y_i - \bar{y})^2}} \qquad (4.30)$$

where

$$\bar{x} = \frac{1}{m} \sum_{i=1}^{m} x_i, \quad \bar{y} = \frac{1}{m} \sum_{i=1}^{m} y_i \qquad (4.31)$$

Another widely adopted linear correlation measurement is root mean squared error (RMSE),

$$RMSE(\mathbf{x}, \mathbf{y}) = \sqrt{\frac{1}{m} \sum_{i=1}^{m} (x_i - y_i)^2} \qquad (4.32)$$

Obviously, if an objective method is perfectly correlated with subjective evaluation, PLCC would be 1 and RMSE be 0; if they are totally unrelated, PLCC would be 0 and RMSE would be infinite. RMSE directly reflected that how much the objective and subjective scores of each sample differ from each other on average, so the scale of it is dependent on that of the scores.

If the nonlinear mapping is tried to be avoided, rank-order correlation coefficient can be applied. The most famous two are Spearman's rank-order correlation coefficient (SROCC) and Kendall's rank-order correlation coefficient (KROCC). SROCC is defined as

$$SROCC(\mathbf{x}, \mathbf{y}) = PLCC(rank(\mathbf{x}), rank(\mathbf{y})) \qquad (4.33)$$

where the function *rank* of a vector returns a vector, in which the i-th item ($i = 1$ to the length of the vector) is the rank of the i-th item in the original vector.

There are three ways to define KROCC; we here only list the most convenient and widely known on,

$$\text{KROCC}(\mathbf{x}, \mathbf{y}) = \frac{C - D}{\frac{1}{2}m(m - 1)} \tag{4.34}$$

where C denotes how many pairs that is consistently correlated between $\mathbf{x}$ and $\mathbf{y}$, and D is the number of the other pairs. Evidently, SROCC and KROCC both equal to 1 when the correlation is perfect and 0 for the worst situation.

Because the nonlinear mapping might work different results for different methods, it is currently more acknowledged to use SROCC as the main metric for evaluating the performance of objective methods. In most cases, the PLCC, RMSE, and SROCC results are all important and cannot replace each other. Of course, there are many other feasible schemes for measurement, but for existing methods, PLCC, RMSE, and SROCC are the most acknowledged metrics to report as the performance of certain objective methods in a database.

References

Bjerhammar, A. (1951). Application of calculus of matrices to method of least squares: With special references to geodetic calculations. Transactions of the Royal Institute of Technology, Stockholm, Sweden: 49.

Breiman, L. (1996). Bagging predictors. *Machine Learning, 24*(2), 123–140.

Casella, G., & Berger, R. L. (2001). *Statistical inference*. Duxbury, MA: Duxbury Press.

Chang, H., Zhang, Q., Wu, Q., & Gan, Y. (2015). Perceptual image quality assessment by independent feature detector. *Neurocomputing, 151*, 1142–1152.

Cormack, L. K. (2005). Computational models of early human vision. In A. C. Bovik (Ed.), *Handbook of image and video processing* (2nd ed.). Amsterdam, Netherlands: Elsevier Academic Press.

Cortes, C., & Vapnik, V. N. (1995). Support vector networks. *Machine Learning, 20*(3), 273–297.

Daly, S. (1992). The visible difference predictor: An algorithm for the assessment of image fidelity. *Proceedings of SPIE, 1616*, 2–15.

Ding, Y., Wang, S., & Zhang, D. (2014). Full-reference image quality assessment using statistical local correlation. *Electronics Letters, 50*(2), 79–80.

Geisler, W. S., & Banks, M. S. (1995). In M. Bass (Ed.), *Handbook of optics*. Manhattan, NY: McGraw-Hill.

Haykin, S. O. (1998). *Neural networks and learning machines*. New York: Pearson Education Inc.

Hinton, G., Osindero, S., & The, Y.-W. (2006). A fast learning algorithm for deep belief nets. *Neural Computation, 18*(7), 1527–1554.

Ho, T. K., Hull, J. J., & Srihari, S. N. (1994). Decision combination in multiple classifier systems. *IEEE Transactions on Pattern Analysis and Machine Intelligence, 16*(1), 66–75.

Kittler, J., Hatef, M., Duin, R., & Matas, J. (1998). On combination classifiers. *IEEE Transactions on Pattern Analysis and Machine Intelligence, 20*(3), 226–239.

Larson, R. J., & Marx, M. L. (2005). *An introduction to mathematical statistics and its applications*. Upper Saddle River, NJ: Prentice Hall Inc.

LeCun, Y. & Bengio, Y. (1995). Convolutional networks for images, speech, and time-series. The handbook of brain theory and neural networks. Cambridge, MA: MIT Press.

LeCun, Y., Bottou, L., Bengio, Y., & Haffner, P. (1998). Gradient-based learning applied to document recognition. *Proceedings of the IEEE, 86*(11), 2278–2324.

Li, Q., & Wang, Z. (2009). Reduced-reference image quality assessment using divisive normalization-based image representation. *IEEE Journal of Selected Topics in Signal Processing, 3*(2), 202–211.

Liu, A., Lin, W., & Narwaria, M. (2012). Image quality assessment based on gradient similarity. *IEEE Transactions on Image Processing, 21*(4), 1500–1512.

Lubin, J. (1993). The use of psychophysical data and models in the analysis of display system performance. In A. B. Watson (Ed.), *Digital images and human vision*. Cambridge, MA: The MIT Press.

Marziliano, P., Dufaux, F., Winkler, S., & Chen, T. (2002). Perceptual blur and ringing metrics: Application to JPEG2000. *Signal Processing: Image Communication, 19*(2), 163–172.

Mood, A. M., Graybill, F. A., & Boes, D. C. (1974). *Introduction to the theory of statistics*. New York: McGraw-Hill Book Company.

Moore, E. H. (1920). On the reciprocal of the general algebraic matrix. *Bulletin of the American Mathematical Society, 26*(9), 394–395.

Moorthy, A. K., & Bovik, A. C. (2010). A two-step framework for constructing blind image quality indices. *IEEE Signal Processing Letters, 17*(5), 513–516.

Moorthy, A. K., & Bovik, A. C. (2011). Blind image quality assessment: From natural scene statistics to perceptual quality. *IEEE Transactions on Image Processing, 20*(12), 3350–3364.

Penrose, R. (1955). A generalized inverse for matrices. *Proceedings of the Cambridge Philosophical Society, 51*, 406–413.

Platt, J. (1998). *Sequential minimal optimization: A fast algorithm for training support vector machines*. (Technical Report MSR-TR-98-14). Microsoft Research.

Portilla, J., Strela, V., Wainwright, M. J., & Simoncelli, E. P. (2003). Image denoising using scale mixtures of Gaussians in the wavelet domain. *IEEE Transactions on Image Processing, 12*(11), 1338–1351.

Rao, K. R., & Yip, P. (1990). *Discrete cosine transform: Algorithms, advantage, applications*. New York: Academic Press.

Saad, M., Bovik, A. C., & Charrier, C. (2012). Blind image quality assessment: A natural scene statistics approach to perceptual quality. *IEEE Transactions on Image Processing, 21*(8), 3339–3352.

Safranek, R. J., & Johnston, J. D. (1989). A perceptual tuned sub-band image coder with image dependent quantization and post-quantization data compression. In *Proceedings of IEEE International Conference on Acoustics, Speech, and Signal Processing* (pp. 1945–1948).

Sheikh, H. R., & Bovik, A. C. (2006). Image information and visual quality. *IEEE Transactions on Image Processing, 15*(2), 430–444.

Sheikh, H. R., Bovik, A. C., & Cormack, L. (2005). No-reference quality assessment using natural scene statistics: JPEG2000. *IEEE Transactions on Image Processing, 14*(11), 1918–1927.

Sheikh, H. R., Sabir, M. F., & Bovik, A. C. (2006). A statistical evaluation of recent full reference image quality algorithms. *IEEE Transactions on Image Processing, 15*(11), 3441–3452.

Sheikh, H. R., Wang, Z., Cormack, L. & Bovik, A. C. (2002). Blind quality assessment for JPEG2000 compressed images. In *Proceedings of IEEE Asilomar Conference on Signals, Systems, and Computers* (pp. 1403–1407).

Simoncelli, E. P., & Olshausen, B. A. (2001). Natural image statistics and neural representation. *Annual Review of Neuroscience, 24*, 1193–1216.

Teo, P. C., & Heeger, D. J. (1994). Perceptual image distortion. *Proceedings of SPIE, 2179*, 127–141.

Wandell, B. A. (1995). *Foundations of vision*. Cary, NC: Sinauer Associates Inc.

Wang, Z., & Bovik, A. C. (2006). *Modern image quality assessment*. San Rafael, CA: Morgan & Claypool.

Wang, Z., & Bovik, A. C. (2009). Mean squared error: Love it or leave it? A new look at signal fidelity measures. *IEEE Signal Processing Magazine, 26*(1), 98–117.

Wang, Z., Bovik, A. C., & Evans, B. L. (2000). Blind measurement of blocking artifacts in images. *Proceedings of IEEE International Conference on Image Processing, 3*, 981–984.

Wang, Z., Sheikh, H. R., & Bovik, A. C. (2002). No-reference perceptual quality assessment of JPEG compressed images. In *Proceedings of IEEE International Conference on Image Processing* (pp. 477–480).

Wang, Z., Wu, G., Sheikh, H. R., Simoncelli, E. P., Yang, E. H., & Bovik, A. C. (2006). Quality-aware images. *IEEE Transactions on Image Processing, 15*(6), 1680–1689.

Wu, H. R., & Yuen, M. (1997). A generalized block-edge impairment metric for video coding. *IEEE Signal Processing Letters, 4*(11), 317–320.

Xu, L., Krzyzak, A., & Suen, C. Y. (1992). Methods of combining multiple classifiers and their applications to handwriting recognition. *IEEE Transactions on Systems, Man, and Cybernetics, 22*(3), 418–435.

Yu, Z., Wu, H. R., Winkler, S., & Chen, T. (2002). Vision-model-based impairment metric to evaluate blocking artifact in digital video. *Proceedings of IEEE, 90*, 154–169.

Zhang, L., Shen, Y., & Li, H. (2014). VSI: A visual saliency-induced index for perceptual image quality assessment. *IEEE Transactions on Image Processing, 23*(10), 4270–4281.

Zhang, L., Zhang, L., Mou, X., & Zhang, D. (2011). FSIM: A feature similarity index for image quality assessment. *IEEE Transactions on Image Processing, 20*(8), 2378–2386.

Chapter 5
Image Quality Assessment Based on Human Visual System Properties

5.1 Introduction

Simulating the behavior of human visual system (HVS) is the target of many research fields in image processing, pattern recognition, and computer vision. Many of the related problems that are very easy for HVS remain largely unsolved for machines, e.g., large-scale object, scene and activity recognition and categorization, vision-based manipulation, face detection and recognition. However, finding how HVS is doing for these problems, i.e., how "biological vision" and "computer vision" interact each other, is a promising approach to seek for solutions and has been proved effective in many specific areas (Dickinson et al. 2009). Image quality assessment (IQA) is no exception. It is safe to state that IQA has benefitted significantly by applying the related findings in the study of HVS structures and properties. After all, it is a natural and reasonable thought to build artificial models of HVS that work similarly to it. Therefore, in this chapter, we are focusing on the objective IQA methods that are designed based on the analysis of HVS.

The responses of HVS on biological level maintain a mystery very much. Knowledge from neuroscience tells us that the only things that neurons do are receiving and sending electrical signals. The neurons in various areas are combined by certain topological structures so that they can be capable to cope with very complicated jobs. The area related to the generation of vision in our brain is called visual cortex. Statistically, 55% of primate neocortex is related to vision (Felleman and Essen 1991), suggesting that vision is undoubtedly the most important sensory of ours. With an extensive number of neurons devoted to the acquisition, processing, 3D reconstruction and many other related jobs for vision, the computational efficiency is extremely high, which makes the violence modeling of HVS almost impossible under current conditions. In addition, it is the cascaded combination of eyes and visual cortex that develops vision, yet the inner structures of eyes are mysterious enough.

© Zhejiang University Press, Hangzhou and Springer-Verlag GmbH Germany 2018
Y. Ding, *Visual Quality Assessment for Natural and Medical Image*,
https://doi.org/10.1007/978-3-662-56497-4_5

Discussions in this chapter are highly related to the introduction and analysis of HVS in Chap. 3. As mentioned in Chap. 3, the hierarchical structure in visual cortex is one of the most important discoveries in vision research in the last few decades (Hubel and Wiesel 1968) and enlightens the computer vision community to build hierarchical models of HVS (Marr 1977; Tenenbaum et al. 2001; Bengio 2009). Concretely, we can build models to simulate each certain area of eye or visual cortex and then synthesize them together for a comprehensive result, instead of regarding HVS as a Black box. The different areas in visual cortex indeed have different interests, so the components for the artificial model of HVS might substantially differ from each other (Kandel et al. 2000). With the development of modern hardware and software, the hierarchical solution is promising for many objects in computer vision. The IQA methods that draw inspirations from the HVS hierarchies will be introduced in Sect. 5.2.

On the other hand, treating HVS as an intact signal processing system and then studying the overall responses is also a feasible solution for IQA. Supposing it is so, HVS would have to be regarded as a highly complex nonlinear system. Nevertheless, many simple properties of HVS have been discovered and can be directly taken advantage of without considering the inner connections between the sub-components of HVS. This approach is sometimes called the "flat" processing scheme, which is generally more efficient than the hierarchical design. The related IQA methods will be discussed in Sect. 5.3.

One specific property widely utilized for IQA methods that we would like to mention in particular is visual attention, in Sect. 5.4. Visual attention is a very important concept of HVS for IQA because it affects subjective image quality perception in quite straightforward way (Peters et al. 2005; Rodríguez-Sánchez et al. 2007). On what degree distortions can be perceived by humans is largely depending on image content. Specifically, distortions in the areas that people are apt to pay more attention to are obviously more noticeable for HVS, while some distortions are even not noticeable for HVS on certain occasions. In practice, visual attention has been taken into account on many IQA methods.

5.2 IQA Based on the Hierarchical Structure

The hierarchical structure of HVS has been introduced in Sect. 3.2. Visual signals are processed through the eyes and the visual cortex in regular sequence. The eyes are basically a pair of band-limit filters that retain certain components of the visual signals. Visual signals are acquired by cells located on the retina called photoreceptors (or photoreceptor cells), including two types: rods and cones, sensitive to the luminance and chromatic information, respectively (Hecht et al. 1942; Baylor et al. 1979; Hurvich 1981). The visual signals obtained by photoreceptors are passed through ganglion cells and lateral geniculate nucleus (LGN) to visual cortex (Kremers 2005).

Visual cortex is divided into various different areas. Generally, the visual signals are preprocessed in the early areas of visual cortex, regions V1, V2, V3, V4, and middle temporal (MT) region (Maunsell and Essen 1983; Orban 2008). Then, the signals are sent to two pathways: ventral pathway, responsible for object recognition (Ungerleider and Mishkin 1982; Tanaka 1996); and dorsal pathway, responsible for space and action detection (Tanaka and Saito 1989; Lewis and Essen 2000; Nakamura et al. 2001). Because IQA is concerned only about the quality perception of images and videos, the areas in charge of object detection, action detection, and other higher-level operations are irrelevant. Therefore, existing methods mainly focus on how HVS processes visual signals through the eye and the early areas of visual cortex.

5.2.1   Cascaded Visual Signal Processing

In the early era of IQA research, many models are constructed by dividing the mission into several sub-problems, assigning them to the corresponding components in HVS, simulating the related components, and combining these simulating models in serial order, as the way the components been simulated are organized in HVS. This way the IQA methods can be regarded as cascaded visual signal processing systems, which try to strictly simulate the hierarchical structure of HVS. IQA methods with this thought can be found at least in (Safranek and Johnston 1989; Daly 1992; Lubin 1993; Teo and Heeger 1994; Lubin 1995; Watson et al. 1997).

We take the visible differences predictor (VDP) as an example for an illustration about how the cascaded visual signal processing methods work (Daly 1992). As the name implies, VDP aims to capture and quantify the degree of visible distortion of an image (i.e., the visible difference between the image and its reference) and regard it as the quality of this image. The essential part of VDP is an HVS model composed by three cascaded stages, as shown in Fig. 5.1, which outputs the probability of detecting the visible distortion of the received image.

At the time that VDP is proposed, the most widely applied objective IQA method would be undoubtedly mean squared error (MSE) (Wang and Bovik 2006). However, MSE has been receiving continuous critics about its inaccuracy, mainly due to its loose connection with the properties of HVS (Girod 1993; Wu and Rao 2005;

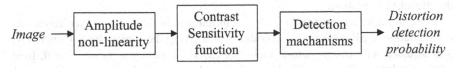

Fig. 5.1 HVS model proposed in Daly (1992)

Wang and Bovik 2009). By constructing models to simulate HVS, what the desire of researchers is to turn the difference between image in pixels into the difference detected by HVS or more correlated with HVS. In other words, because humans do not perceive images as the simple combination of pixels, what HVS is sensitive to is not explicitly reflected by image pixels. This fact concerns the change from pixel variations to human visual sensitivity variations.

As claimed by the authors, their HVS model addresses three main sensitivity variations, light level, spatial frequency, and signal content, respectively, corresponding to the three cascaded stages shown in Fig. 5.1. The light adaptive properties of the retina lead to nonlinearity in the light level, accounted by HVS in the amplitude nonlinearity computation stage. The spatial frequency variation is considered by the contrast sensitivity function (CSF), which is actually a combinational effect by the optical effects of the eyes, the neural circuitry, etc. Image content variations are related to higher-level properties about the post-receptor neural circuitry in HVS and are concluded as masking.

The sensitivity S is defined as the inverse of the minimum contrast level C that is sufficient to produce a threshold response,

$$S = \frac{1}{C} \tag{5.1}$$

in which C is determined by the Michelson definition,

$$C = (L_{max} - L_{mean})/L_{mean} \tag{5.2}$$

where L_{max} and L_{mean} are the maximum and mean luminance level of the image, respectively.

The amplitude nonlinearity denotes that there is a nonlinear function between the perceptive lightness and the value of luminance. Therefore, a nonlinear mapping function based on grayscale images is adopted (Sezan et al. 1987), which is defined for each separate pixel. The normalized response $\mathbf{R}$ of the mapping of each location (x, y) for grayscale image $\mathbf{I}$ can be represented as

$$\mathbf{R}(x, y) = \frac{\mathbf{I}(x, y)}{\left(\mathbf{I}(x, y) + (C_1 \mathbf{I}(x, y))^{C_2}\right)} \tag{5.3}$$

where C_1, C_2 are parameters to be set. The effect of the mapping is similar to that of the cubic root (Mannos and Sakrison 1974; McCann et al. 1976), but is adaptive to the global illumination.

CSF, an important property of HVS as introduced in Sect. 3.2, is then taken into consideration. CSF is taken into account by modifying the sensitivity S. Concretely, S is adjusted with many factors, including radial frequency ρ, orientation θ, light adaptation level l, image size s, lens accommodation related to viewing distance d, and eccentricity e,

$$S(\rho, \theta, l, s, d, e) = P \cdot \min \left\{ S'\left(\frac{\rho}{\mathrm{bw}_d \cdot \mathrm{bw}_e \cdot \mathrm{bw}_\theta}, l, s \right), S'(\rho, l, s) \right\} \tag{5.4}$$

where the inner function S' is defined as

$$S'(\rho, l, s) = \left(\left(3.23 \left(\rho^2 s \right)^{-0.3} \right)^5 + 1 \right)^{-1/5} \cdot 0.9 A_l \rho e^{-0.9 B_l \rho} \sqrt{1 + 0.06 e^{0.9 B_l \rho}} \tag{5.5}$$

The parameters bw_d, bw_e, bw_θ, A_l, and B_l are

$$\mathrm{bw}_d = 0.856 d^{0.14} \tag{5.6}$$

$$\mathrm{bw}_e = \frac{1}{1 + 0.24e} \tag{5.7}$$

$$\mathrm{bw}_\theta = 0.15 \cos(4\theta) + 0.85 \tag{5.8}$$

$$A_l = 0.801(1 + 0.7/l)^{-0.2} \tag{5.9}$$

$$B_l = 0.3(1 + (100/l))^{0.15} \tag{5.10}$$

Actually, constructing CSF model is a common approach for IQA and other computer vision applications. Aside from Daly's works, a few CSF models can be found in (Mannos and Sakrison 1974; Wilson and Bergen 1979; Watson 1987a; Watson and Ahumanda 2005).

The third and final stage of the model is called detection mechanism, which mainly aims to deal with image content-related effects that can be concluded as masking (see Sect. 3.3). Evidence of the masking property is found in the field from both neurophysiology (Hubel and Wiesel 1962; De Valois et al. 1982) and psychophysics (Blackmore and Campbell 1969; Stromeyer and Julesz 1972), indicating that HVS is selective in spatial frequency. The selectivity of spatial frequency is modeled by the cortex transform (Watson 1987b), operated as a cortex filter that is the cascaded combination of filters separately accounting for the radial frequency selectivity (*rad*) and the orientation selectivity (*ori*),

$$cortex(\rho, \theta) = rad(\rho) \cdot ori(\theta) \tag{5.11}$$

where ρ and θ are the amplitude and angle in the Fourier plane. The radial frequency filter is the difference of two low-pass filters (*lp*). The low-pass filter is set with a transition region with width w modeled by a Hanning window,

$$lp(\rho) = \begin{cases} 1 & |\rho < \rho_{0.5} - 0.5w \\ \frac{1}{2}\left(1 + \cos \frac{\pi(\rho - \rho_{0.5} + 0.5w)}{w}\right) & |\rho_{0.5} - 0.5w < \rho < \rho_{0.5} + 0.5w \\ 0 & |\rho > \rho_{0.5} + 0.5w \end{cases} \tag{5.12}$$

and

$$rad(\rho) = lp(\rho)\big|_{\rho_{0.5}=2^{-(k-1)}} - lp(\rho)\big|_{\rho_{0.5}=2^{-k}} \tag{5.13}$$

And the orientation filter is defined as

$$ori(\theta) = \begin{cases} \frac{1}{2}\left(1 + \cos\left(\frac{\pi|\theta-\theta_c(l)|}{\theta_w}\right)\right) & \|\theta - \theta_c(l)\| \leq \theta_w \\ 0 & \|\theta - \theta_c(l)\| > \theta_w \end{cases} \tag{5.14}$$

where

$$\theta_c(l) = (l - 1) \cdot \theta_w - \frac{\pi}{2} \tag{5.15}$$

$$\theta_w = \frac{\pi}{L} \tag{5.16}$$

Adjusting the integer parameters k and l ($k = 1$ to K, $l = 1$ to L) can construct a series of filters with different bandwidths and angles. Let $\mathbf{T}_{k,l}$ and $\mathbf{R}_{k,l}$ be the filtered tested and reference images, and the contrast difference of them is defined as

$$\Delta\mathbf{C}_{k,l}(x,y) = \mathbf{C}_{k,l}^{\mathbf{T}}(x,y) - \mathbf{C}_{k,l}^{\mathbf{R}}(x,y) = \frac{\mathbf{T}_{k,l}(x,y)}{\overline{\mathbf{T}_K}} - \frac{\mathbf{R}_{k,l}(x,y)}{\overline{\mathbf{R}_K}} \tag{5.17}$$

The normalized mask contrast $\mathbf{M}$ for parameter k and l is determined with the help of the CSF and the cortex filter,

$$\mathbf{M}_{k,l}(x,y) = \mathcal{F}^{-1}\big(\mathcal{F}(\mathbf{R}) \cdot csf(u,v) \cdot cortex_{k,l}(u,v)\big) \tag{5.18}$$

with which a "threshold evaluation image" $\mathbf{Th}$ is constructed,

$$\mathbf{Th}_{k,l}(x,y) = \left(1 + \big(k_1\big(k_2|\mathbf{M}_{k,l}(x,y)|\big)^s\big)^b\right)^{1/b} \tag{5.19}$$

where parameters k_1, k_2, s, and b are to be set empirically.

Finally, the probability of distortion detection can be calculated for each pixel position (x, y). On a specific sub-band (k, l), the detection probability is

$$\mathbf{P}_{k,l}(x,y) = 1 - \exp\left(-\left(\frac{\Delta\mathbf{C}_{k,l}(x,y)}{\mathbf{Th}_{k,l}(x,y)}\right)^\beta\right) \tag{5.20}$$

By synthesizing all sub-bands, detection probability for (x, y) is

$$\mathbf{P}(x,y)=1 - \prod_{k=1}^{K}\prod_{l=1}^{L}\mathbf{P}_{k,l}(x,y) \qquad (5.21)$$

And the overall quality of the image can be derived with the **P** value for all pixels.

The above introduced objective IQA method is a good example of cascaded algorithm structure. It is capable to take into account the important HVS properties such as light adaptation, CSF, masking, and the components are organized in the order similar to the functionalities of HVS parts. However, several flaws are evident, which are very common for this type of works. Firstly, it neglects higher-level quality-aware features in visual cortex such as edge, shape, texture, and so on, while mainly focuses on simple features related to the retina and early vision. Certainly, this can be fixed by building more complex models, but that would get the following two problems more severe. The second problem is that there exists approximation for each stage of the model. With more stages cascaded, error caused by the approximation would be accumulated and leads to huger disparity with the real HVS system. Last by not least, the computational time of a cascaded system is generally larger than a parallel one. In fact, the cascaded structure is less common in recent years, mostly due to the efficiency problem, and parallel feature extraction (Sect. 5.2.2) and the "Black box" structures (Sect. 5.3) are more and more adopted in more recent years.

5.2.2   Parallel Feature Extraction

Modern IQA methods tend to favor parallel structures (Wang et al. 2004; Zhang et al. 2011; Chang et al. 2015). Several quality scores derived by multiple paths can be easily mapped to the final quality score through various means (see Sect. 4.3). Superficially, this is in contrast with the hierarchical structure of HVS that transmits visual signals layer by layer. But it is actually feasible because with the knowledge about the properties and interests of each part of HVS, current feature extraction tools are capable to capture the information that a certain component of HVS is sensitive to without prior knowledge about the outputs of its previous components. Therefore, it is very important that we know what properties are appealing to different areas of HVS. Fortunately, the related progress is significant in the last few decades (Krüger et al. 2013).

An example we would like to introduce is (Ding et al. 2017b), which we will refer to as multifeature extraction and synthesis (MFES) for convenience. As the authors summarize, the parallel methods for IQA should follow: All features are quality-aware; different features should aim for different aspects to make a quality description as comprehensive as possible; duplicated efforts should be as few as possible. Good news is that the work division of HVS is quite clear. So, by modeling different parts of HVS with different schemes, a comprehensive description can be quite straightforwardly developed while maintains concise. The part of HVS at earlier stage of visual signal processing is sensitive to lower-level features. Concretely, the photoreceptors on the retina merely sense the lights with different luminance and color, but the visual signals are then assembled as edges, bars and spatial frequencies in V1 and to textures and shapes in V2 (Hyvärinen et al. 2009; Willmore et al. 2010). Later regions in visual cortex are responsible for more higher-level jobs such as object and action detection and are not concerned too much for image quality perception.

The framework of MFES is given in Fig. 5.2. Aiming for capturing features correlating with the properties of V1 and V2, several tools are employed. Specifically, Sobel operators are utilized to extract edges from images, log Gabor filter bank for the spatial frequency information, and local patterns are extracted to describe textures. Disparities of the features between of the tested and reference images are computed as the quality indices and are mapped to the final quality score by support vector regression (SVR).

Sobel operator (Sobel 1970) is one of the most popular edge detection tools in the last few decades, which is normally considered superior than other operators such as Roberts' (1965) and Prewitt's (1970) because consideration is given to all

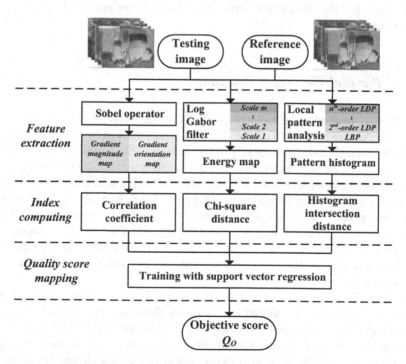

Fig. 5.2 Framework of MFES

eight neighbors of a pixel and their relative distances. Conventionally, a Sobel operator is actually a pair of masks $\mathbf{S_H}$ and $\mathbf{S_V}$ extracting the horizontal and vertical edges, respectively,

$$\mathbf{S_H} = \begin{bmatrix} -1 & 0 & 1 \\ -2 & 0 & 2 \\ -1 & 0 & 1 \end{bmatrix}, \quad \mathbf{S_V} = \begin{bmatrix} -1 & -2 & -1 \\ 0 & 0 & 0 \\ 1 & 2 & 1 \end{bmatrix} \quad (5.22)$$

Thus, the horizontal and vertical gradient maps $\mathbf{G_H}$, $\mathbf{G_V}$ of a grayscale image $\mathbf{I}$ are

$$\mathbf{G_H} = \mathbf{I} * \mathbf{S_H}, \quad \mathbf{G_V} = \mathbf{I} * \mathbf{S_V} \quad (5.23)$$

A common approach to make use of the horizontal and vertical gradients is to combine them as the gradient magnitude map $\mathbf{G_M}$ and gradient orientation map $\mathbf{G_O}$,

$$\mathbf{G_M}(x,y) = \sqrt{(\mathbf{G_H}(x,y))^2 + (\mathbf{G_V}(x,y))^2} \quad (5.24)$$

$$\mathbf{G_O}(x,y) = \arctan \frac{\mathbf{G_V}(x,y)}{\mathbf{G_H}(x,y)} \quad (5.25)$$

Log Gabor filter (Field 1987) is a variation of Gabor filter (see Sect. 5.3) with the modification that the frequency responses are represented on log axis, so that the cortical cells are better simulated. The frequency response of 2D log Gabor filter in polar form is

$$\mathbf{LG}_{\sigma_f,\sigma_\theta}(f, \theta) = \exp\left(-\frac{(\log(f/f_0))^2}{2(\log(\sigma_f/f_0))^2} - \frac{(\theta - \theta_0)^2}{2\sigma_\theta^2}\right) \quad (5.26)$$

where f_0 and θ_0 are the central frequency and orientation of the filter, and σ_f and σ_θ are the scale and angle bandwidth.

Detailed discussion about frequency responses and Gabor filters is given in Sect. 5.3, so we are not going too deep about the log Gabor responses. An obvious thing is that from neither the characteristic of the frequency responses or as the exponential form in the above equation implies the items in LG is in the complex form. As usual, we are interested in the mode of the response, which are also referred to as the local energies.

Moreover, a set of different frequencies and orientations creates a filter bank with different scales s and orientations θ, deriving the energy maps $\mathbf{E}_{s,\theta}$. The authors adopt a four-scale and four-orientation filter bank, i.e., s = integers 1 to 4, $\theta = 0$, 0.25π, 0.5π, and 0.75π. On each scale, the four spatial frequency maps of different orientations are summed up and averaged. The final energy map on scale s is

$$E_s(x, y) = \frac{1}{4} \sum_{i=0}^{3} E_{s,0.25\pi i}(x, y) \tag{5.27}$$

The edge and local frequency features are both low level. To exploit higher-level features, especially to be attractive to V2, higher-level information about shapes and textures is captured, using local pattern analysis. Concretely, local binary pattern (LBP) (Ojala et al. 1996, 2002) and local derivative pattern (LDP) (Tan and Triggs 2010) are utilized to extract from the first to third-order patterns. The term "order" here denotes the times of derivative computation to be operated.

LBP is a successful feature extraction tool in pattern and texture recognition. The most initial version of LBP is to represent a pixel as an eight-bit binary code, each bit denoting the relative values of the pixel and one of its neighbors. The binary code C of a pixel p is

$$C(p) = \sum_{i=0}^{7} \mathrm{sgn}(p, p_i) \cdot 2^i \tag{5.28}$$

where p_i (i = integers 0 to 7) are the eight neighboring pixels of p and function sgn can be defined as

$$\mathrm{sgn}(x, y) = \begin{cases} 1, & x > y \\ 0, & x \le y \end{cases} \tag{5.29}$$

The obvious weakness of this simplest LBP is that the code is determined for a pixel merely related to its neighbors with Chebyshev distance 1. This factor is easily noted and is tried to be revised in later versions of LBP and other local pattern extracting tools, including LDP. Therefore, for more comprehensive texture extraction, LDP is employed. LDP calculates the derivates of pixels between pixels in an iterative manner. For instance, the second-order LDP computes the relative pixel values between the central pixel p and the pixels with Chebyshev distance 1 from p; the pixels with Chebyshev distance 1 and those with Chebyshev distance 2 from p. Then, the two relations are utilized for further process.

Actually, LBP and its variants (including LDP) are commonly adopted for texture analysis in objective IQA (Zhang et al. 2013a, b; Du et al. 2016; Freitas et al. 2016; Ding et al. 2017a). More detailed introduction about local patterns is given in Sect. 6.3 as we have classified the texture analysis into the natural scene statistics (NSS) model category.

The authors then employ three different means to measure the disparities between the tested and reference images for the three features. The fact that different approaches are applied suggests that it is an open problem. A set of commonly adopted different measurements are introduced in Sect. 4.3. Generally, the selection from them is experiment-oriented, so is the selection of mapping strategy, to some degree. Also, they are not affecting the basic principles of the method, so they are in fact quite irrelevant here.

In the very beginning, we have discussed the complementary nature of the features in MFES from the prospective of neural cortex, what we have not mentioned is that from the view of implementation. Obviously, both LBP and LDP are only concentrating on the relative values of pixels, but neglect how much they are different from each other. While Sobel operators and log Gabor filters directly exploit the difference between pixels, the effect of log Gabor filters is quite like a giant sharpening mask, and Sobel operator is operated only on the first order, so they make good supplements as well. The reported experimental results also demonstrate this, illustrated by a much better performance of the combinational method than any feature implemented alone.

Moreover, an exciting fact is that this performance improvement costs very little extra time consuming. By extracting each feature separately in parallel, the consuming time is only equal to the time spent for extracting the slowest feature plus the time for mapping the final quality score that is usually negligible. Thus, computational time is very easy to control. In the specific case of MFES, the feature slowest in extraction is LDP, because higher-order LDP has to be computed based on the results of the lower order, and that is why the authors stop at the third order. Nevertheless, by adjusting this parameter, the computational burden is kept under control. The performance gain and consuming time invariance are true for other parallel structure based IQA methods, explaining its popularity.

5.3 IQA Based on Responses of Human Visual System

5.3.1 Contrast Sensitivity and Multiple Scales

The contrast sensitivity function (CSF) has been brought about in Chap. 3 and previous sections in this chapter many times. An intuitive understanding is that the HVS is more sensitive a certain degree of contrast than the others. The existence of this property makes HVS like a band-pass filter in the frequency domain. Therefore, given an image, if the image resolution and the distance between the image and viewers are known, we can find out what spatial frequency is the most appealing to HVS. However, a common assumption is that the distance between the image and viewers is changeable for the latter, so the distortion on a large range of frequencies is ought to be taken into account. This concept is referred to the multiscale property of HVS.

The multiscale property is considered not only in the field of IQA, but in many other related fields in computer vision as well. For one instance, in object detection, the purpose is usually to detect the desired objects on all possible scales. Similarly, we are expecting IQA methods to successfully notice the distortions on all possible scales. The multiscale structure similarity (MS-SSIM) is one among the earliest and best-known method making use of this property and is thus taken as an example for us to explain how it is assisting IQA.

Actually, SSIM is regarded as an NSS-based method, and Chap. 6 of this book will give a more detailed introduction about it. The idea of SSIM is to decompose an image into three channels, luminance l, contrast c, and structure s, all of which are reflecting image quality and are orthotropic to each other. The exponential multiplication of them is the quality score Q of the image,

$$Q = l^\alpha \cdot c^\beta \cdot s^\gamma \tag{5.30}$$

where parameters α, β, and γ are to be learned, defining the relative importance of the three features. Because l, c, and s are derived for each pixel with its neighbors in a moving mask, the scale of the extracted features is fixed.

The framework of MS-SSIM is given in Fig. 5.3. The images, both the tested image and its reference, are transmitted to a set of images with different scales using a bank of M low-pass filters down-sampling images with a factor of 2. On each scale, the contrast and structure indices are computed, while the luminance is only computed for the largest scale (from the image after $M - 1$ times of low-pass filtering), which is because the luminance is a scale-irrelevant index in this case, and only finding the luminance on the smallest images is very computationally efficient. Finally, the indices are mapped to quality score Q also by exponential multiplication,

$$Q = l_M^{\alpha_M} \cdot \prod_{i=1}^{M} \left(c_i^{\beta_i} \cdot s_i^{\gamma_i} \right) \tag{5.31}$$

where the exponential parameters are set to determine weights for the scales. Specifically, there are settled with the help of subjective experiments (Faugeras and Pratt 1980; Gagalowicz 1981; Heeger and Bergen 1995; Portilla and Simoncelli 2000). Furthermore, to simplify parameter selection and to balance the parameters, it is ensured that

$$\alpha_i = \beta_i = \gamma_i, \quad \forall i = 1, 2, \ldots, M \tag{5.32}$$

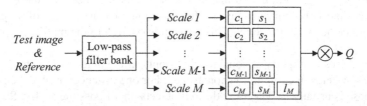

Fig. 5.3 Framework of MS-SSIM

$$\sum_{i=1}^{M} \gamma_i = 1 \qquad (5.33)$$

Experimental results reported by the authors have demonstrated the effectiveness of introducing the multiscale property. With M growing from 1, there is evident performance gain of single-scale SSIM; while when M becomes too large, the performance is falling. This fact corresponds to the band-pass property of CSF very well. By combining the scales, the predicting accuracy is further improved.

The abovementioned low-pass filter bank is in most cases the same to the concept of Gaussian pyramid, except that the latter is much more flexible with its variations. But of course, the low-pass filter bank is not the only way to construct a multiscale structure. At the very least, Laplacian pyramid and steerable pyramid are also favored in image processing and computer vision (Burt and Adelson 1983; Manduchi et al. 1997), and there are extensive studies upon different smoothing kernels to construct image pyramids other than the Gaussian mask (Burt 1981; Crowley 1981; Crowley and Parker 1984; Crowley and Sanderson 1987; Meer et al. 1987). These pyramid construction strategies are similar that iterative filtering is included. A more direct way is to compute the multifrequency response of the images, such as wavelet transform and Gabor filters, as discussed in the following.

5.3.2 Wavelet Transforms and Gabor Filters

To simulate CSF, a straightforward idea is by band-pass filtering in the frequency domain. Fourier transform is the basic way to convert signals to frequency domain. For a grayscale image, it follows traditional 2D Fourier transform,

$$F(\omega_x, \omega_y) = \frac{1}{2\pi} \iint f(x,y) e^{-j(\omega_x y + \omega_y y)} \mathrm{d}x \mathrm{d}y \qquad (5.34)$$

where ω_x and ω_y are the frequency along the two dimensions, and F is the transformed function in the frequency domain. In practice, we are always dealing with digital visual signals, so discrete Fourier transform (DFT) is really been processed. For convenience, we will only show the DFT format of 1D signals. Assuming x (n) is a finite sequence with length N, then its DFT is

$$X(k) = \sum_{n=0}^{N-1} x(n) e^{-j\frac{2\pi}{N}kn} \qquad (5.35)$$

Actually, half of the DFT results are capable of carrying all necessary information, so discrete cosine transform (DCT) is also widely adopted for image processing. By utilizing the fast Fourier transform (FFT) (Cooley and Tukey 1965), the computational complexity of DFT and DCT is decreased from n^2 to $n \log n$, therefore is very popular for operation on modern computational systems.

The major problem of Fourier transform and the mentioned variations of it are that the spatial information is totally neglected in the signals transferred to the frequency domain, yet the spatial relationships between pixels are rather important for visual signals. In other words, they are only good at processing stationary signals but not nonstationary ones, while signals are almost always nonstationary in reality. Fortunately, there are two different kinds of approaches aiming to fix the problem, both of which have been studied very thoroughly in the last few decades, namely wavelet transform and short-time Fourier transform (STFT). Moreover, because they are frequency extraction tools, they are by definition able to well capture the multiscale features.

A formal representation of 1D continuous wavelet transform of function f is

$$W_f(a,b) = \frac{1}{\sqrt{|a|}} \int_{-\infty}^{+\infty} f(x) \overline{\varphi\left(\frac{x-b}{a}\right)} \, dx \qquad (5.36)$$

where function φ is referred to as the wavelet basis, wavelet kernel, or wavelet function. It is easy to notice that the difference between wavelet transform and the Fourier transform is that the kernel φ is defined by two parameters, a and b, denoted as scaling factor and displacement factor, rather than the fixed kernel $\exp(-jwt)$ that has only one parameter, frequency. The wavelet transform of a 1D signals is 2D, along the axis of scale a and displacement b. The responses are located on spatial domain by combing a and b. In addition, by using different kernels, the results of wavelet transform can be very different. Popular wave functions include Harr, Daubechies, Mexican hat, Morlet, Meyer.

Instead of replacing the kernel, STFT is developed from Fourier transform by adding a window function g,

$$F(\omega, b) = \int f(x,y) g^*(x-b) e^{-j\omega x} dx \qquad (5.37)$$

where g^* is the conjugate function of g. The effect of windowing is to intercept the original function and remain the part within the window. If multiple windows are adopted with the same shape but different sizes, the responses of different scales are generated. Obviously, the selection of specific window function affects the responses of STFT. Gaussian function is the most favorable window in STFT. The 2D STFT adopting Gaussian window is the famous Gabor filter, the response of which is

$$G(x,y; \lambda, \theta, \psi, \sigma, \gamma) = \exp\left(-\frac{x'^2 + \gamma^2 y'^2}{2\sigma^2}\right) \exp\left(j\left(2\pi \frac{x'}{\lambda} + \psi\right)\right) \qquad (5.38)$$

where λ, θ, ψ, σ, and γ are parameters for the Gaussian function, representing the wavelength of the sinusoidal factor, the orientation of the normal to the parallel stripes, the phase offset, the standard deviation of the kernel, and the spatial ratio, respectively, and x' and y' are the pixels coordinates modified for better representation,

$$x' = x \cos \theta + y \sin \theta \qquad (5.39)$$

$$y' = -x \sin \theta + y \cos \theta \qquad (5.40)$$

The width of the window for STFT, also called the support length, determines the discrimination of the signals in the spatial and frequency domain. With infinite support length, STFT degenerates to Fourier transform, so the spatial discriminating capability is totally lost. With support length decreasing, spatial discriminating capability would increase, at the cost of frequency discrimination. Therefore, it is impossible to achieve both the best spatial and frequency discrimination. This fact is very important in signal processing, known as Heisenberg's uncertainty principle or Heisenberg's indeterminacy principle. Employing Gaussian function as the window gives the best trade-off between them. On the other hand, it has been proved that Gabor response is highly correlated with the response of HVS. Therefore, Gabor filters are attached great importance in image processing and computer vision (Raju et al. 2004; Pati and Ramakrishnan 2008; Gdyczynski et al. 2014). STFT other than Gabor filter is utilized very rarely, thus it will be neglected in following discussions.

Because wavelet transform and Gabor filter are both very popular for image processing, Gabor filter is sometimes referred to as Gabor wavelet. Obviously, they are by definition different from each other, but with respect to the filtering results, they do share a lot in common. Most importantly, they are both able to extract the spatial frequency features from images, which are very attractive to HVS. However, there also exists huge difference between them. The wavelets achieve multiresolution automatically. On the contrary, although the multiscale property can be modeled by Gabor filters, the required parameters have to be determined manually. For a selected set of parameters, the response of the Gabor filter is only single-scale and single-orientation. The common employment of Gabor filters is to construct a filter bank, containing filters with different scales and orientations, so that a comprehensive response can be obtained. Usually, the scales are set as less than ten and cover the frequency that the peak of CSF corresponds to; the orientations are determined as several values (preferably 4 or 6) that distribute evenly in the range of 0 to π.

Due to the importance of frequency analysis for visual signals, a large amount of works that making use of the frequency responses can be found in the literature of IQA. Concretely, all of the abovementioned tools can be employed for IQA, including DFT (Narwaria et al. 2012), DCT (Brandao and Queluz 2008; Zhai et al. 2008; Saad et al. 2010, 2012; Ma et al. 2011; Shen et al. 2011; Gao et al. 2013; Ma et al. 2013, 2016; Xue et al. 2014; Liu et al. 2015; Wang et al. 2015), wavelet transforms (Sheikh et al. 2005a; Lu et al. 2008, 2010; Gao et al. 2009; Li and Wang 2009; Li et al. 2009, 2011; Liu and Yang 2009; Soundararajan and Bovik 2012; Rezazadeh and Coulombe 2013; Ding et al. 2014a; Liu et al. 2014a, b; Gu et al. 2015; Wu et al. 2016), Gabor filters and Gabor filters' variations (such as log Gabor filters introduced in Sect. 5.2) (D'Angelo et al. 2010; Larson and Chandler 2010; Ye and Doermann 2012; Zhang and Chandler 2013; Ding et al. 2014b, 2017b; Hong et al. 2014; Qian et al. 2014; Jiang et al. 2015; Zhao et al. 2016).

5.3.3 Another Prospective Based on Information Theory

Treating HVS as a packaged signal processing system, the properties of it can be described either by finding its response, as discussed in above, or studying the information contents it carries. In other words, the information theory offers us another prospective to analysis related problems for visual signal processing. IQA has also been inspired by this point of view, with evident dated back to a long time ago. As mentioned, the most ancient measurement, MSE, is defined as

$$\text{MSE}(\mathbf{T}, \mathbf{R}) = \frac{1}{N} \sum_{i=1}^{N} \left(\mathbf{T}_i^2 - \mathbf{R}_i^2 \right) \tag{5.41}$$

where $\mathbf{T}$ and $\mathbf{R}$ are the tested image and its reference, and N is the total pixel number. MSE is simply the Euclidean distance between the two images averaged along all dimensions. With a nonlinear modification, the metric peak signal-to-noise ratio (PSNR) is constructed,

$$\text{PSNR}(\mathbf{T}, \mathbf{R}) = 10 \log_{10} \left(\frac{(2^b - 1)^2}{\text{MSE}(\mathbf{T}, \mathbf{R})} \right) \tag{5.42}$$

where b is the sampling bits of the signal, which is 8 in most modern image processing systems.

The original purposed of this transformation is to use the logarithm axis to better to let the quality scores more similar to HVS responses. From a more recent opinion, this modification for predicting accuracy is irrelevant because it does not affect the monotonicity of the quality prediction, so the performance of MSE and PSNR for IQA is basically the same.

However, an important contribution of PSNR is that the concept of information is introduced. By definition, PSNR means to measure the ratio of the correct signal to the noise, the term "peak" denotes the maximum MSE, $(2^b - 1)$, equaling 255 mostly. Obviously, an infinite PNSR denotes the perfect image quality, because the denominator, the noise, is 0. The noise given by MSE is directly related to the difference between the tested and reference images.

Another way to put signal-to-noise ratio (SNR) can be the information fidelity of the distorted image or the mutual information shared by two images. For two well-known examples, information fidelity criterion (IFC) (Sheikh et al. 2005b) and visual information fidelity (VIF) (Sheikh and Bovik 2006) are based on this idea, both of which have achieved way better results comparing to MSE and PSNR. Similar ideas can be found in the literature. Visual signal-to-noise ratio (VSNR) proposed in (Chandler and Hemami 2007) is one instance that modifies PSNR to better simulate HVS and will be discussed in the following.

Since the ratio of signal to noise is to be computed, how to quantify noise is firstly determined. Unlike MSE, the authors employ L1 distance to define the noise $\mathbf{N}$,

$$\mathbf{N} = \mathbf{T} - \mathbf{R} \tag{5.43}$$

which straightforwardly reflect the distortion contained in $\mathbf{T}$. The root-mean-square (RMS) contrast C of a visual signal is defined as (taking $\mathbf{N}$ as an example) (Moulden et al. 1990; Tiippana et al. 1994; Kingdom et al. 1995; Bex and Makous 2002)

$$C(\mathbf{N}) = \frac{1}{\mu_{\mathbf{L(R)}}} \sqrt{\frac{1}{N} \sum_{i=1}^{N} \left(L(\mathbf{N}_i + \mu_{\mathbf{R}}) - \mu_{\mathbf{L(N}_i + \mu_{\mathbf{R}})} \right)^2} \tag{5.44}$$

where L is a nonlinear modification for image pixel $\mathbf{N}_i$ according to (Poynton 1998), and

$$\mu_{\mathbf{R}} = \frac{1}{N} \sum_{i=1}^{N} \mathbf{R}_i \tag{5.45}$$

$$\mu_{\mathbf{L(R)}} = \frac{1}{N} \sum_{i=1}^{N} L(\mathbf{R}_i) \tag{5.46}$$

$$\mu_{\mathbf{L(N}_i + \mu_R)} = \frac{1}{N} \sum_{i=1}^{N} L(\mathbf{N}_i + \mu_{\mathbf{R}}) \tag{5.47}$$

For a certain spatial frequency f, the specific contrast signal-to-noise ratio (CSNR), CSNR_f, is

$$\text{CSNR}_f(\mathbf{R}, \mathbf{N}) = \frac{C(\mathbf{R}_f)}{C(\mathbf{N}_f)} \tag{5.48}$$

An attempt then made to model the property of HVS is to set a threshold that defines whether the distortion is visible. The threshold CT is closely related to image contrast,

$$CT(N_f|T) = \frac{C(\mathbf{R}_f)}{CSNR_f^{thr}}$$

(5.49)

where $CSNR_f^{thr}$ is a constant uniquely defined by f,

$$CSNR_f^{thr} = a_0 f^{a_2 \ln f + a_1}$$

(5.50)

with parameters a_0, a_1, and a_2 determined by previous studies (Ramos and Hemami 2001). Based on $CSNR_f^{thr}$, the global-precedence-preserving CSNR with noise signal $\mathbf{N}$ is developed as

$$CSNR_f^*(\mathbf{N}) = b_0(\mathbf{N}) f^{b_2(\mathbf{N}) \ln f + b_1(\mathbf{N})}$$

(5.51)

where

$$b_0(\mathbf{N}) = -a_0 v(\mathbf{N}) + a_0$$

(5.52)

$$b_1(\mathbf{N}) = (1 - a_1) v(\mathbf{N}) + a_1$$

(5.53)

$$b_2(\mathbf{N}) = (-1 - a_2) v(\mathbf{N}) + a_2$$

(5.54)

Index $v(\mathbf{E})$ is an index of visibility, ranging in [0, 1]. The concept $CSNR_f^*$ extends the definition of $CSNR_f^{thr}$ adaptively that the threshold is reflected partly by the noise carried.

CSNR is utilized mainly to decide whether the distortions in $\mathbf{T}$ are visible. More formally, for $\mathbf{N}$ at center frequency f_m, if the perceived contrast $C(\mathbf{N}_{f_m})$ is below threshold $CT(\mathbf{N}_{f_m}|\mathbf{R})$, the distortion is deemed invisible. Signals at different frequencies are decomposed from an M-level separable discrete wavelet transform, and parameter m is an integer from 1 to M.

The perceived visual distortion (VD) is a linear combination of the perceived contrast (pc) of $\mathbf{N}$ and the global-precedence-preserving (gp) CSNR that are, respectively, defined as

$$d_{pc} = C(\mathbf{N})$$

(5.55)

$$d_{gp} = \sqrt{\sum_{m=1}^{M} \left(C^*(\mathbf{N}_{f_m}) - C(\mathbf{N}_{f_m}) \right)^2}$$

(5.56)

VD is represented as

$$VD = \alpha d_{\text{pc}} + (1 - \alpha) \frac{d_{\text{gp}}}{\sqrt{2}} \tag{5.57}$$

VSNR is then uniquely determined by VD and the contrast of the original image $C(\mathbf{I})$, using the form of SNR computation in information theory. A simple modification based on (5.42) can do the trick. Replacing the "peak" and the MSE between $\mathbf{T}$ and $\mathbf{R}$ with $C(\mathbf{I})$ and the square of VD, respectively, and VSNR is represented as

$$VSNR(\mathbf{T}, \mathbf{R}) = 10 \log_{10} \left(\frac{C(\mathbf{I})^2}{VD^2} \right) \tag{5.58}$$

with expansion

$$VSNR(\mathbf{T}, \mathbf{R}) = 10 \log_{10} \frac{C(\mathbf{I})^2}{\left(\alpha d_{\text{pc}} + (1 - \alpha) \frac{d_{\text{gp}}}{\sqrt{2}} \right)^2} \tag{5.59}$$

5.4   IQA Based on Visual Attention

5.4.1   Visual Saliency Models

Visual attention is one of the most important mechanisms of HVS, which refers to a cognitive process that enables HVS pay more attention to more attractive regions of the visual scene. The visual attention mechanism can be roughly classified into two categories, top-down and bottom-up ones. The former is driven by task and deal with high-level cognitive factors, taking advantage of new discoveries in other fields, e.g., psychology, neurobiology, and computer vision. While the latter is driven by stimulus, based on the hypothesis that visual attention is caused by low-level visual stimulus, e.g., luminance, color, edge, orientation. Most of the existing visual attention models belong to the latter category, since bottom-up attention is more thoroughly studied and the current knowledge about visual attention is quite limited that how human attention affects the perception of image quality is yet to be clearly understood (Toet 2011).

Usually, visual attention is also called visual saliency in computer vision, and mainly refers to bottom-up processes. The computed visual saliency (or conspicuity) for an image is often a two-dimensional topographic representation, a saliency map, with each location corresponding to the image using a scalar quantity. With higher saliency level, the location attracts more attention of HVS. Visual saliency is intrinsically related to IQA, since distortions in the regions that are more appealing tend to have a larger impact on subjective sensation than that are less attractive (Li et al. 2015). Studies suggest that integrating visual saliency into IQA is beneficial for performance improvement in terms of theoretical, experimental, and practical perspectives (Engelke et al. 2011; Liu and Heynderickx 2011; Wang et al. 2016).

Almost all saliency detection models are directly or indirectly inspired by cognitive properties, since the saliency detection results are supposed to agree with human perception. There exist a number of cognitive saliency detection models that have strong bindings to cognitive findings. The earliest notable and perhaps the most well-known saliency model is (Itti et al. 1998), which is designed based on the neuronal architecture of the primates' early visual system. It is the first implementation of the visual saliency computational architecture introduced in (Koch and Ullman 1985). This model defines saliency as distinctiveness of an image region to its surroundings and combines multiscale image features into a single topographical saliency map. Saliency toolbox proposed in (Walther and Koch 2006) is an extension to the Itti model by a process of inferring the extent of a proto-object at the attended location from the maps that are used to compute the saliency map. The adaptive whitening saliency (AWS) model is based on the idea of decorrelation of neural responses (Garcia-Diaz et al. 2012). (Le Meur et al. 2006) models the behavior of HVS by taking CSF, visual masking, and the center-surround interactions into consideration. The spatio-temporal saliency model (Marat et al. 2009) predicts eye movement during video free viewing, which is inspired by the biology of the first steps of the HVS that two signals corresponding to the two main outputs of the retina are extracted from the video stream and then split into elementary feature maps that used to form saliency maps.

Graphical models are another class of saliency detection models, which are based on graph theory and conform to probabilistic framework. This kind of models can be regarded as generalization of the Bayesian theory. Among them, a typical graph-based visual saliency (GBVS) model (Harel et al. 2007) is accomplished in two steps, first generating activation maps using extracted feature vectors and then normalizing them in a way which highlights conspicuity and admits combination with other maps. Suppose that a feature map M is obtained from the input image $\mathbf{I}$, then the dissimilarity between two nodes on it, (i, j) and (m, n), is defined as

$$d((i,j),(m,n)) = \left| \log \frac{\mathbf{I}(i,j)}{\mathbf{I}(m,n)} \right| \tag{5.60}$$

Assuming a fully-connected directed graph, $\mathbf{G_A}$, is formed that connects every node in I with all the others. A weight is assigned for the edge from node (i, j) to node (m, n),

$$w_1((i,j),(m,n)) = d((i,j),(m,n)) \exp\left(-\frac{(i-m)^2 + (j-n)^2}{2\sigma^2} \right) \tag{5.61}$$

where σ is a free parameter. Obviously, the weight of the directed edge is proportional to their dissimilarity and closeness. Then, a Markov chain is defined on $\mathbf{G_A}$, and the equilibrium distribution of this chain is calculated as the activation map, which naturally accumulate mass at nodes that have high dissimilarity with their surrounding nodes. The next normalization step is kind like a mass-concentration procedure. Given an activation map $\mathbf{A}$, then every node is connected to all other nodes to construct graph $\mathbf{G_N}$. Then, the edge from (i, j) to (m, n) is weighted by

$$w_2((i,j),(m,n)) = \mathbf{A}(m,n)\exp\left(-\frac{(i-m)^2 + (j-n)^2}{2\sigma^2}\right) \qquad (5.62)$$

Similarly, the weights of the outbound edges are normalized and the resulting graph is treated a Markov chain. Then, the equilibrium distribution can be calculated as saliency map. The resulting graphs are treated as Markov chains by normalizing the weights of the outbound edges of each node to 1 and by defining an equivalence relation between nodes and states, as well as between edge weights and transition probabilities. Their equilibrium distribution is adopted as the activation and saliency maps.

There exist other similar methods. The E-saliency model (Avraham and Lindenbaum 2010) introduces a validated stochastic model to estimate the probability that an image region is attractive to HVS, and this probability is defined as saliency. This algorithm starts with a rough preattentive segmentation and then uses a graphical model approximation to select the optimal segments that are more likely to be attractive.

Bayesian saliency models are based on probabilistic formulation, in which prior knowledge (scene context) and sensory information (extracted features) are probabilistically combined according to Bayes rule. This kind of saliency models takes advantage of the statistics of natural scenes or other features that attract attention. Torralba proposed a Bayesian framework for visual search as well as saliency detection (Torralba 2003). A typical saliency model using natural statistics (SUN) is proposed in (Zhang et al. 2008a, b), which is a Bayesian framework from which bottom-up saliency emerges naturally as the self-information of visual features, and overall saliency emerges as the pointwise mutual information between the features and the target being searched. In this model, the measure of saliency is derived from natural image statistics rather than the statistics of the particular images being viewed. In (Bruce and Tsotsos 2009), the authors propose the attention-based information maximization (AIM) model by taking advantage of Shannon's self-information to transform the image feature plane into a dimension that closely corresponds to visual saliency. The formula for saliency detection in SUN is similar to that in (Torralba 2003) and (Bruce and Tsotsos 2009) because they both based on self-information, motivated by the hypothesis that a foreground object is likely to have conspicuous features comparing with the background. In (Itti and Baldi 2009), a Bayesian definition of surprise is introduced to capture subjective aspects of sensory information, where the surprising stimuli are defined as differences that significantly change beliefs of an observer and measure how data affects an observer by computing the dissimilarity between posterior and prior beliefs.

Frequency domain models have also been studied by researchers for years. For example, by analyzing the log-spectrum of the input image, a simple yet efficient Fourier transform-based method is proposed in (Hou and Zhang 2007) to extract the spectral residual (SR) of the image in spectral domain and construct the corresponding saliency map in spatial domain. The SR model has a very low computational complexity and is independent of features, categories, or other forms of prior knowledge of the objects. Given an image $\mathbf{I}$, its frequency domain spectra $\mathbf{F_I}$ can be obtained by Fourier transformation, and then amplitude spectrum $\mathbf{A}$ and phase spectrum $\mathbf{P}$ can be calculated as

$$\mathbf{A(I)} = \mathrm{Re}(\mathbf{F_I}) \tag{5.63}$$

$$\mathbf{P(I)} = \mathrm{Im}(\mathbf{F_I}) \tag{5.64}$$

where Re and Im denote the real and imaginary parts, respectively. Define the spectral residual $\mathbf{R}$ as

$$\mathbf{R(I)} = \log(\mathbf{A(I)}) - \log(\mathbf{A(I)}) * \mathbf{H} \tag{5.65}$$

where $\mathbf{H}$ is a local average filter, and * is the convolution symbol. Then, the SR saliency map $\mathbf{SR}$ is obtained by

$$\mathbf{SR(I)} = \mathbf{G} * \left(\mathcal{F}^{-1}(\exp(\mathbf{R(I)} + \mathbf{P(I)})) \right)^2 \tag{5.66}$$

where $\mathbf{G}$ is a Gaussian filter to smooth the saliency map for better visual effect.

Following work of Hou et al., a phase spectrum Fourier transform (PFT) model is proposed (Guo et al. 2008). In this paper, the authors point out that it is the phase spectrum of the Fourier transform, rather than the amplitude spectrum used in SR, that is the key in calculating the location of salient regions. In their later work, the phase spectrum of quaternion Fourier transform (PQFT) model is proposed to calculate the spatio-temporal saliency map of an image using its quaternion representation (Guo and Zhang 2010). The additional motion dimension allows the phase spectrum to represent spatio-temporal saliency in order to perform attention selection not only for images but also for videos. In addition, the PQFT model can calculate the saliency map of an image under various resolutions. Other approaches to make use of Fourier transform for saliency detection can be found in (Achanta et al. 2009) where a frequency-tuned salient region detection model is proposed to obtain full resolution saliency maps with well-defined boundaries of salient object. By extracting low-level features of color and luminance, the model is computational efficient and easy to implement. In (Bian and Zhang 2009), a refined spectral domain based spatio-temporal saliency map model is proposed based on the idea of spectral whitening (SW). In this model, the whitening process is used as a normalization procedure in the construction of a map that only represents salient features while suppressing redundant background information.

Other saliency models that do not fall into above categories also deserve to be mentioned here. By combining prior knowledge from three aspects, a novel conceptually simple yet efficient saliency model called saliency detection by combining simple priors (SDSP) is proposed in (Zhang et al. 2013a, b). Firstly, it is assumed that the procedure of salient object detection of HVS can be modeled by band-pass filtering. Secondly, HVS are likely to pay more attention to the center of an image. Thirdly, warm colors are more attractive to human eyes than cold colors. In (Liu et al. 2014a, b), the authors propose a novel saliency detection framework termed as saliency tree. In this model, the original image is firstly decomposed into primitive regions, and then the regional saliency of each region is obtained by integrating global contrast, spatial sparsity, and object prior of primitive regions to build a reasonable basis for generating the saliency tree. After that, a saliency-directed region merging approach is proposed to generate the saliency tree. Finally, the pixel-wise saliency map is derived through a systematic saliency tree analysis process. Zhu et al. introduce a robust background measure method that characterizes the spatial layout of image boundaries (Zhu et al. 2014). Moreover, they proposed a principled optimization framework to integrate multiple low-level features, including the background measure, to obtain clean and uniform saliency maps.

Beyond those saliency models mentioned above, numerous of saliency models are proposed and more detailed reviews can be referred to (Borji and Itti 2013; Zhang et al. 2016). Figure 5.4 shows the saliency maps generated by some of the typical models mentioned above for a certain image selected from LIVE image quality assessment (IQA) database (Sheikh et al. 2006). The involved models include SR, SDSP, GBVS, and SUN.

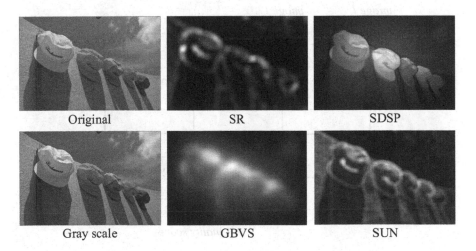

Fig. 5.4 Saliency maps generated by four models for an image in LIVE database

5.4.2   Application of Visual Saliency in IQA

Over the past decade, the close relationship between visual saliency and quality perception has attracted more and more attention from researchers. It is proved that IQA metrics can be improved by incorporating visual data (Larson and Chandler 2008). Based on the hypothesis that distortion occurring in regions that attract more attention is more annoying than in other regions, efforts have been made to incorporate computational VS models into IQA.

Visual saliency models can be integrated into IQA in two ways. The first and the more widely adopted way is to use saliency map as a weighting function in quality score pooling stage (Saha and Wu 2013; Ma and Zhang 2008; Moorthy and Bovik 2009; Tong et al. 2010; Farias and Akamine 2012; Ma et al. 2010). On the other hand, visual saliency map can also serve as feature maps representing the image's local quality (Zhang and Li 2012; Zhang et al. 2014a, b).

For a reference image $\mathbf{I_r}$ and a distorted image $\mathbf{I_d}$, suppose that the local quality map $\mathbf{S}$ is calculated by quantify similarity between feature maps $\mathbf{FM_r}$ and $\mathbf{FM_d}$ derived from $\mathbf{I_r}$ and $\mathbf{I_d}$. Let $\mathbf{VS}$ denotes the obtained visual saliency map, and then the first integration scheme based methods share a general form represented as

$$\mathbf{S}'(i,j) = \frac{\mathbf{S}(i,j) \cdot \mathbf{VS}(i,j)}{\sum_{\mathbf{I}} \mathbf{VS}(i,j)} \tag{5.67}$$

where $\mathbf{S}'$ is the quality map after the integration. The framework of this weighting strategy is illustrated in Fig. 5.5.

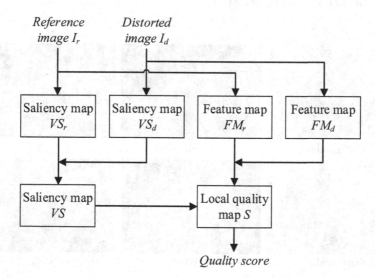

Fig. 5.5 Framework of IQA methods integrating visual saliency maps with quality maps

Note that the visual saliency map **VS** can be computed according to either the reference image (**VS$_r$**) or the distorted image (**VS$_d$**). Some specific methods make use of both **VS$_r$** and **VS$_d$**, e.g., using the maximum value of saliency in each location.

Methods belonging to the second category combine the saliency maps with the feature maps instead of the quality maps. The feature map **FM** (either **FM$_r$** or **FM$_d$**) is integrated with **VS** (correspondingly **VS$_r$** or **VS$_d$**) in the form

$$\mathbf{FM}'(i,j) = \mathbf{FM}(i,j) \cdot \mathbf{VS}(i,j) \tag{5.68}$$

The computed weighted feature map **FM'** is used in subsequent processing. Framework of IQA methods in this kind of integration strategy is shown in Fig. 5.6.

The abovementioned two classes of integrating methods are both pixel-wise schemes. In (Lin et al. 2013), a patch-based combination scheme is proposed. Saliency maps computed by GBVS are incorporated with three classical full-reference IQA metrics, SSIM (Wang et al. 2004), and FSIM/FSIM$_C$ (Zhang et al. 2011). Take SSIM as an example. Given an image **I**, let **SSIM** denotes the obtained quality map and **VS** denotes the saliency map, and then the quality map after integration with **VS** is represented as

$$\mathbf{SSIM}'(i,j) = \frac{(\mathbf{SSIM}(i,j))^{\alpha} \cdot (\mathbf{VS}(i,j))^{\beta}}{\sum_{\mathbf{I}} (\mathbf{VS}(i,j))^{\beta}} \tag{5.69}$$

where

$$\alpha = K_{\mathrm{SSIM}} \cdot \overline{\mathbf{SSIM}} \tag{5.70}$$

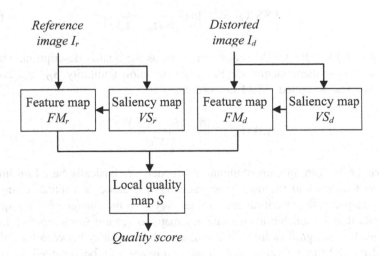

Fig. 5.6 Framework of IQA methods integrating visual saliency maps with feature maps

$$\beta = K_{\text{VS}} \cdot \text{corr}(\mathbf{VS_r}, \mathbf{VS_d}) \tag{5.71}$$

Parameters K_{SSIM} and K_{VS} are constants. The variations of parameters α and β adjust the contribution of local SSIM values and lead to the results that the importance of SSIM and VS is changing over image regions. Indices FSIM and FSIM_C are also modified in this way.

The application of VS in IQA is not limited only within the aforementioned ways. A weighted full-reference IQA method proposed in (Wen et al. 2017) is an example. For the reference image and distorted image $\mathbf{I_r}$ and $\mathbf{I_d}$, $\mathbf{VS_r}$ and $\mathbf{VS_d}$ are the VS maps derived with SR saliency model. To enhance the contrast between the salient and nonsalient regions, a simple transform is introduced,

$$\mathbf{VS}'(i,j) = \begin{cases} \lambda \mathbf{VS}(i,j), & \mathbf{VS}(i,j) \geq t \\ 1, & \text{otherwise} \end{cases} \tag{5.72}$$

where λ is a constant ($\lambda > 1$) for image enhancement, and t is a threshold determined as

$$t = \frac{3}{N} \sum_{\mathbf{I}} \mathbf{VS} \tag{5.73}$$

where N is the total pixel number of $\mathbf{I}$. Then, a weight matrix $\mathbf{W}$ is calculated,

$$\mathbf{W} = \max(\mathbf{VS}'_\mathbf{r}, \mathbf{VS}'_\mathbf{d}) \tag{5.74}$$

With the weight matrix, PSNR can be improved to a weighted PSNR that better predicts image quality,

$$\mathbf{PSNR_W} = 20 \log \frac{(2^b - 1)N}{\|\mathbf{W}(\mathbf{I_d} - \mathbf{I_r})\|^2} \tag{5.75}$$

Another IQA index GSIM is also employed as the quality description, which is composed of gradient magnitude $\mathbf{S_G}$, edge direction similarity $\mathbf{S_D}$, and contrast matrix $\mathbf{S_C}$. The weighted GSIM is represented as

$$\mathbf{GSIM_W} = \frac{\|\mathbf{S_G} \cdot \mathbf{S_D} \cdot \mathbf{S_C} \cdot \mathbf{W}\|^2}{\|\mathbf{W}\|^2} \tag{5.76}$$

Since the bottom-up computational VS models are basically based on image's low-level features and the quality degradation will cause perceptible changes in saliency maps, so the visual saliency values vary with the change of image quality. Therefore, it is reasonable to use saliency maps as feature maps representing the image quality straightforwardly. For example, the similarity between the saliency maps derived from reference and distorted images can be regarded as quality descriptor (Zhang et al. 2014a, b; Zhang and Li 2012). The framework of an IQA metric where saliency maps serve as feature maps is shown in Fig. 5.7.

In (Zhang and Li 2012), the authors propose an IQA metric named spectral residual-based similarity (SR-SIM), which exhibits high-quality predicting accuracy as well as low computational cost. The saliency model employed in this algorithm is the previously introduced SR model (Hou and Zhang 2007), which plays a dual role in the algorithm. On one hand, the SR saliency map is exploited as a weighting function in pooling stage, as discussed above. On another hand, it also acts as a feature map characterizing the image's local quality and used to compute the local similarity map between the reference and distorted images. The similarity between the computed SR maps $\mathbf{SR_d}$ and $\mathbf{SR_r}$ is

$$\mathbf{SIM} = \frac{2\mathbf{SR_r} \cdot \mathbf{SR_d} + C}{\mathbf{SR_r} + \mathbf{SR_d} + C} \tag{5.77}$$

where C is a positive constant to avoid the denominator being zero for stability improvement.

Another state-of-the-art VS-based full-reference IQA method is the visual saliency index (VSI) proposed in (Zhang et al. 2014a, b). For a distorted image $\mathbf{I_d}$ and its reference $\mathbf{I_r}$, the visual saliency maps $\mathbf{V}$, gradient feature maps $\mathbf{G}$, and chrominance feature maps $\mathbf{M}$ and $\mathbf{N}$ are calculated, for both $\mathbf{I_r}$ and $\mathbf{I_d}$. In this paper, the saliency maps are computed with multiple schemes. The similarity between saliency maps and that between gradient maps are defined in similar ways,

$$\mathbf{SIM_V}(i,j) = \frac{2\mathbf{V_r}(i,j) \cdot \mathbf{V_d}(i,j) + C_1}{\mathbf{V_r}(i,j)^2 + \mathbf{V_d}(i,j)^2 + C_1} \tag{5.78}$$

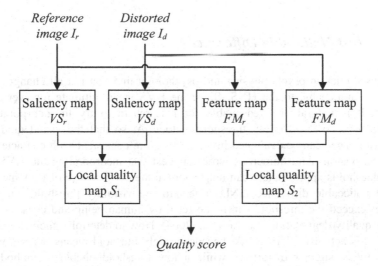

Fig. 5.7 Framework of IQA metric where saliency maps serve as feature maps

$$\mathbf{SIM_G}(i,j) = \frac{2\mathbf{G_r}(i,j) \cdot \mathbf{G_d}(i,j) + C_2}{\mathbf{G_r}(i,j)^2 + \mathbf{G_d}(i,j)^2 + C_2} \tag{5.79}$$

where C_1 and C_2 are positive constants and so is C_3 in the next equation. The chrominance similarity is the product of that of both chrominance channels,

$$\mathbf{SIM_C}(i,j) = \frac{2\mathbf{M_r}(i,j) \cdot \mathbf{M_d}(i,j) + C_3}{\mathbf{M_r}(i,j)^2 + \mathbf{M_d}(i,j)^2 + C_3} \cdot \frac{2\mathbf{N_r}(i,j) \cdot \mathbf{N_d}(i,j) + C_3}{\mathbf{N_r}(i,j)^2 + \mathbf{N_d}(i,j)^2 + C_3} \tag{5.80}$$

Then, the local similarity is defined as the weighted product of these features,

$$\mathbf{SIM}(i,j) = \mathbf{SIM_V}(i,j) \cdot \mathbf{SIM_G}(i,j)^{\alpha} \cdot \mathbf{SIM_C}(i,j)^{\beta} \tag{5.81}$$

where α and β are to be learned to define the importance of the features.

In this method, visual saliency is merely utilized as quality features so far. A weight matrix is also derived with the saliency maps,

$$\mathbf{W}(i,j) = \max(\mathbf{V_r}(i,j), \mathbf{V_d}(i,j)) \tag{5.82}$$

The final VSI is

$$\mathrm{VSI} = \frac{\sum \mathbf{SIM} \cdot \mathbf{W}}{\sum \mathbf{W}} \tag{5.83}$$

The reported performance of VSI has demonstrated the effectiveness of the visual saliency to act both as a quality-aware feature and a weight definer in the pooling stage.

5.4.3 Just Noticeable Difference

Previous studies in psychophysics and psychology find that not all changes in an image are noticeable to the HVS. Since the HVS is the ultimate receiver of the images, it is natural and reasonable for researchers to try to incorporate the knowledge of HVS visibility thresholds into IQA, so that the visual quality of distorted images can be evaluated in a manner more consistent with characteristics of human vision. And numerous works suggest that incorporating the HVS visibility thresholds appropriately can help improve the performance of IQA methods.

Just noticeable difference (JND) refers to the visibility threshold, and only changes exceed the threshold can be detected by human being and cause the perceptual quality degradation (Legras et al. 2004). How to determine the value of this threshold is actually a hard work, because a high threshold means insensitivity to the relatively slighter distortion, while a low threshold unable to embody the significance of JND. This task is related to the HVS characteristics and is adaptive to the contents in the visual field. The concept of JND can be traced back to the Weber–Fechner law (Boring 1942), which can be expressed as

$$\frac{\Delta \mathbf{I}}{\mathbf{I}} = K \tag{5.84}$$

where $\mathbf{I}$ is the background luminance, $\Delta \mathbf{I}$ is the just noticeable luminance variation over the background, and K is a constant called Weber fraction.

Most of the existing computational models for JND (Hahn and Mathews 1998; Chou and Li 1995; Chiu and Berger 1999; Ramasubramanian et al. 1999) are mainly based on the effect of the basic stimuli described below. In single-stimulus JND experiments, a test image $\mathbf{I_t}$ can be represented as,

$$\mathbf{I_t} = \mathbf{I_o} + \mathbf{s} \tag{5.85}$$

where $\mathbf{I_o}$ is the original undistorted image, and $\mathbf{s}$ is visual stimulus.

If $\mathbf{s}$ is with changing contrast, for example, a sine-wave, then the JND threshold also affected by spatial and temporal frequencies of s, which is known as the CSF (Barten 1999). Specifically, spatial frequencies reveal the changes in spatial domain, while temporal frequencies reflect the change of visual scene over the time. The psychophysical experiments show that the contrast threshold decreases in extra low and extra high spatial/temporal frequencies.

It is known that the responses of HVS to intensity variation are nonlinear. Considering a patch in an image surrounded by the background, the JND is affected not only by the difference itself but also by background's intensity. Weber–Fechner law suggests that the just noticeable $\mathbf{s}$ increases with the increasing of background luminance $\mathbf{I_o}$. Weber–Fechner law is maintained over a wide range of background luminance but not valid at very low or very high luminance conditions, where the JND increases remarkably.

In case of JND experiments with real-world images, the test image $\mathbf{I_t}$ can be represented as

$$\mathbf{I_t} = \mathbf{I_o} + b(\mathbf{I_d} - \mathbf{I_o}) \tag{5.86}$$

where $\mathbf{I_o}$ is the original image, $\mathbf{I_d}$ is the distorted version of $\mathbf{I_o}$, b is the weighting coefficient to be fitted. Observers are asked to compare $\mathbf{I_t}$ and $\mathbf{I_o}$ with an increasing b, then the JND can be defined as a value of b, which makes majority of the observers able to tell the difference between them.

According to the domain for the JND threshold being calculated, the existing computational models can be roughly classified into two categories, sub-band-based models and pixel-based models.

For models of the former category, an image is firstly transferred into a sub-band domain before all the subsequent operations. The most widely utilized transformation is the DCT decomposition (Hahn and Mathews 1998; Jia et al. 2006; Zhang et al. 2008a, b; Wei and Ngan 2009), since it has been used in many compression standards. And the pixel-based models are also applied in many application, such as motion estimation (Yang et al. 2003a), quality evaluation (Lin et al. 2003), etc.

The major factors that affect the JND mechanism are CSF, luminance adaptation, intra-band masking, and inter-band masking. Thus, the JND can be calculated as the product of the base threshold related to the spatial CSF and other elevation parameters result from other three factors. Given an image $\mathbf{I}$ of size $P \times Q$, let i denotes the position of an $N \times N$ DCT block in $\mathbf{I}$ ($i = 0, 1, 2, ..., PQ/N^2 - 1$), ($m$, n) denotes a DCT sub-band (m, $n = 0, 1, 2, ..., N - 1$). Then, the DCT-based JND can be expressed by

$$\text{JND}_{\text{DCT}}(i, m, n) = T_{\text{CSF}}(i, m, n) \prod_j \eta_j(i, m, n) \qquad (5.87)$$

where T_{CSF} (i, m, n) denotes the base threshold related to the spatial CSF, $\eta_j(i, m, n)$ represents the parameter result from a factor, and j is the ordinal of the factors.

The spatial CSF describes the influence of spatial frequency to HVS. Let L be the background luminance, and the relation between contrast sensitivity S and threshold T is defined as

$$S = \frac{L}{T} \qquad (5.88)$$

The contrast sensitivity S is a function of L and spatial frequency f. The model proposed in (Ahumada and Peterson 1992) is based on the experimental data in (van Nes and Bouman 1967). For a specific L level, there are downward parabolic relations between S and f (Barten 1999). Things are slightly different when it comes to spatial CSF, the vertex and the steepness of parabola depend on L.

The measured experimental data in (van Nes and Bouman 1967) can be fitted to the parabola equation on a logarithmic scale as presented in (Ahumada and Peterson 1992),

$$\log S = \log S_{\max} - K \left(\log f - \log f_p \right)^2 \qquad (5.89)$$

or

$$\log T = \log T_{\min} + K \left(\log f - \log f_p \right)^2 \qquad (5.90)$$

The two above equations have the same essence. Positive constant K determines the steepness of the parabola, and f_p is the vertex of the parabola. The empirically solution of the equation is also given in (Ahumada and Peterson 1992).

But when (5.80) is used for the (m, n)-th DCT sub-band, some modification is required. Firstly, the spatial summation effect should be taken into consideration, since an invisible change in a sub-band may be noticed if presented together with changes from other sub-bands. Therefore, a coefficient b ($0 < b < 1$) is introduced.

Secondly, the oblique effect should also be reduced, since the experimental data involved in (5.80) is only vertical spatial frequencies. For an arbitrary sub-band, the threshold is slightly higher, so another coefficient r is introduced for the absolute threshold of (m, n)-th DCT sub-band of the i-th block,

$$\log T'(i,m,n) = \log \frac{T_{\min}}{r + (1 - r)\cos^2 \theta(m,n)} + K(i)\big(\log f(m,n) - \log f_p(i)\big)^2$$

(5.91)

where

$$f(m,n) = \frac{1}{2N}\sqrt{\left(\frac{m}{w_x}\right)^2 + \left(\frac{n}{w_y}\right)^2}$$

(5.92)

$$\theta(m,n) = \arcsin \frac{2f(m,0)f(0,n)}{f^2(m,n)}$$

(5.93)

Parameters w_x and w_y are the horizontal and vertical visual angles of a pixel, respectively, calculated by

$$w = 2\arctan \frac{\delta}{2d}$$

(5.94)

where δ (either along the orientation of x or y, and so is w) is the display width of a pixel, d is the viewer distance.

Then, the base threshold result from spatial CSF can be calculated as

$$T_{\text{CSF}}(i,m,n) = \frac{G}{\lambda_m \lambda_n (L_{\max} - L_{\min})} T'(i,m,n)$$

(5.95)

where $L_{\max}$ and $L_{\min}$ are the maximum and minimum luminance, G is the maximum number of gray levels, for example, G equals 256 for 8-bit binary representation. And the DCT normalizing coefficient λ is defined as

$$\lambda_u = \begin{cases} \sqrt{\frac{1}{N}}, & u = 0 \\ \sqrt{\frac{2}{N}}, & u \neq 0 \end{cases}$$

(5.96)

As to the pixel-wise JND models, the JND is directly calculated on the original visual content without transformation to sub-band beforehand, so it is more convenient and cost-effective to implement. Numerous pixel-wise JND models have been developed (Ramasubramanian et al. 1999; Yang et al. 2003a; Uzair and Dony 2017) and applied in situations where easier operations are performed in spatial domain, such as motion estimation (Yang et al. 2003b), decoded image quality assessment (Lin et al. 2003), etc.

For pixel-wise JND models in spatial domain, two major factors are mainly considered, namely luminance adaption and texture masking. Visibility threshold is relevant to background luminance as described above. The texture masking effect refers to the fact that complex texture can reduce the visibility of changes, so that changes in textured regions are more likely to be noticed than in smooth regions. The two factors always exist simultaneously. For an image $\mathbf{I}$ of size $P \times Q$, let (i, j) denotes the position of a pixel ($i = 0, 1,..., P - 1, j = 0, 1,..., Q - 1$). Then, a nonlinear additivity model can be used to determine the visibility threshold T (Yang et al. 2003a),

$$T(i,j) = T_l(i,j) + T_t(i,j) - C(i,j)\min\{T_l(i,j), T_t(i,j)\} \qquad (5.97)$$

where T_l and T_t denote the visibility threshold for luminance adaption and texture masking, and C accounts for the overlapping masking effect ($0 < C \leq 1$).

In (Chou and Li 1995), the authors approximately describe the luminance adaption as

$$T_l(i,j) = \begin{cases} 17\left(1 - \sqrt{\frac{L(i,j)}{127}}\right) + 3, & L(i,j) \leq 127 \\ \frac{3}{128}\left(L(i,j) - 127\right) + 3, & \text{otherwise} \end{cases} \qquad (5.98)$$

where $L(i, j)$ is the background luminance of (i, j), calculated by the average luminance with a small neighborhood. The texture masking effect is presented as

$$T_t(i,j) = \mu \cdot g(i,j) \cdot e(i,j) \qquad (5.99)$$

where μ is a control parameter, g is the weighted average of gradient around (i, j), and e is an edge-related weight of (i, j).

Due to the validity of measuring the visual visibility for image distortion, it is no doubt that incorporating JND is beneficial to many image processing applications. For instance, researchers have been trying to integrate JND into quality prediction to improve the performance of existing algorithms (Corporation 1997; Lin et al. 2003). JND-based perceptual metrics can also be applied to other situations, for example, image synthesis. A typical framework of integrating JND in IQA is given in Fig. 5.8, which shows that in short, JND is introduced to guide the computation of the quality map. For one specific instance in IQA, a full-reference IQA metric taking advantage of JND is proposed in (Toprak and Yalman 2017), where the JND threshold is determined by using edge and texture knowledge of great importance in terms of the HVS's psychophysical characteristics. In this paper, the distorted images are transformed into YUV channels and separated into small blocks, and then the JND threshold values of edge and texture version are determined by using calculated background luminance value T_l, edge knowledge T_e, and texture knowledge T_t,

Fig. 5.8 A typical framework integrating JND for IQA

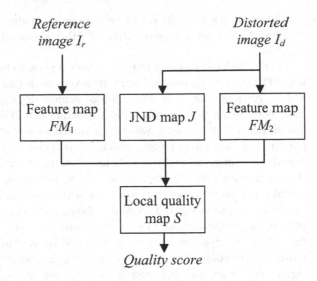

$$\mathrm{JND}_e = T_e + \frac{0.5T_1}{T_e} \tag{5.100}$$

$$\mathrm{JND}_t = T_t + \frac{0.5T_l}{T_t} \tag{5.101}$$

Then, the original and distorted image blocks are compared, taking the JND values into consideration, to measure the quality degradation of distorted image.

Although there is still room for JND models to be improved, it is sufficient for current knowledge upon it to serve for purposed in image processing and computer vision, IQA included. With the development of novel JND models, the integration of them into IQA should be more commonly witnessed. Unlike VS, JND is seemingly hard to serve as quality-aware feature by itself, yet it is capable to offer assistance for almost all IQA frameworks.

5.5 Summary

This chapter has discussed the properties of HVS that that are or potentially are useful for designing IQA methods, and selected examples of successful IQA methods making use of HVS characteristics are introduced to highlight the effectiveness of this designing approach. In the history of the last few decades, discoveries in visual neuroscience and the applications in engineering including image processing and computer vision have influenced, interacted, and inspired each other. The findings in neuroscience has given lots of novel ideas of IQA study as well as proofs for the theoretical ground of IQA methods. Although it has to be admitted that HVS is still yet to be thoroughly studied, we can safely state that studying IQA from the prospective of HVS sensation has become completely feasible.

Firstly, the basic structure of HVS is known to us. In the past, researchers used to take HVS as a Black box and study its properties regardless of its inner structure. This is certainly due to the limitation of neurobiological findings. Nowadays, we have been aware that the topological structure that HVS has is basically hierarchical, rather than flat, as it was once assumed, or any other form. The hierarchies exist in both the visual cortex, which is the region in cortex in charge of visual signal processing, and the regions before that, which are composed by the eyes and the connections between the eyes and the visual cortex. According to this hierarchical structure, we know that our ability with vision, including the ability to judge image quality, to detect and recognize things and faces, to track the movement of objects, etc., is the accumulation of processing through multiple stages. Concretely, the cells in our eyes are categorized into two kinds, those to sense luminance and those to sense color. The cells are operating independently upon each other, yet the signals they carry are integrated and edges, bars, and spatial frequencies are perceived by simple cells in the first visual cortex. Later in the second visual cortex, shapes and textures are composed. In further stages, or in the visual cortex regions that are further from the visual signal receptors, the visual signals are gradually more and more abstracted. Finally, we can let the simple luminance and color signals compose objects and see if they match certain things in our memory (recognition) and detect and tract their locations (localization). HVS is a highly mature system that has been evolved to be very effective for visual signal processing, so, it is reasonable to make use of its hierarchical structure to accomplish the compute vision tasks that are easy for human. Actually, the hierarchical structure is employed for human neural system for other purposes other than generating vision. The success and popularization of neural network is a good example of modeling this hierarchy. Focusing on IQA, the modeling can be simplified because low-level features that are attractive for the eyes and early visual cortex regions are enough for quality description, unlike highly abstracted goals such as detection, recognition, etc. Therefore, simulating HVS by constructing cascaded procedures to model its different stages can be feasible for IQA. Problem is the present understanding upon HVS is still too shallow to give accurate simulation for each part of it, even only for the early vision. For a cascaded system, the error caused by incorrect modeling is likely to accumulate to generate disasters. Therefore, another thought is inspired to model each part of HVS and combine the results in parallel. This way is practical also because that modeling a stage of HVS is not necessarily requiring the responses of its prior stages. Generally, the parallel framework is more accurate and computationally faster, and the regression models produced by machine learning tools further evoke its development.

Secondly, HVS can be studied as an intact system. In other words, it can be treated as a Black box. There are two ways to make use of the Black box, both by finding the relationship between its outputs and inputs. The first approach is to compute the responses of it. HVS is capable to decompose visual signals according to different spatial frequencies, and its sensitivity to visual signals varies according to the frequencies. This decomposition can be modeled simply by frequency analysis. Because Fourier transform abandons the spatial information, other frequency analysis tools are generated, among which two most widely used are wavelet transform and short-time Fourier transform. Analyzing signals at different frequencies is also referred to as analysis at different scales. As the term implies, higher scale denotes lower frequency. So, the ability of frequency decomposition is also called the multiscale property. Wavelet transform and short-time Fourier transform (STFT) are two commonly used tools to deal with the multiscale property of HVS. Wavelet transforms adopt kernels to adaptively achieve the multiscale functionality, while STFT adopts window functions for the same purpose in a manual manner, to some extent. Using different kernels or windows, the computed responses can be very different, so the selection among specific tools is a tricky task. Fortunately, studies in neurobiology have shown that using Gaussian window for STFT can generate responses that are very similar to the actual responses of HVS. STFT with Gaussian windows is called Gabor filter. Gabor filters can be set with different scales and orientations by setting different parameters, so that the multiscale representation is constructed. With the decomposed visual signals, we can concentrate differently on responses at different frequencies. The sensitivity of HVS achieves maximum at a certain frequency and drops with either rising or falling of the frequency, making HVS like a band-pass filter. This property is called contrast sensitivity function (CSF), where contrast can be regarded as another term to refer to the scale or frequency. CSF has received a lot of attention, and lots of mathematical models are constructed to simulate it. When analyzing the spatial frequencies, it is practical to refer to CSF to learn the favors of HVS. For instance, we can endow weights for different frequencies according to how much the specific frequency is appealing to HVS. Actually, most of modern IQA methods have taken the multiscale property of HVS into account, because distortions may occur on various scales as well, and the multiscale model can be effective in assisting the IQA methods, not to mention the ability of the frequency analyzing tools to extract quality-aware features, among which the applications of discrete cosine transforms, wavelet transforms, and Gabor filters are extremely common. The second approach is based on information theory, to find out how the visual information is lost with distortions. Peak signal-to-noise ratio (PSNR) is based on the thought of information content. Problem of PSNR is that the measured information is not necessarily related to human sensation upon image quality, so the definition of information is then modified, often referred to as "visual information," to suggest the relationship between the information and visual quality. In this sense, the definitions of signals and noises are also modified accordingly. Also, there are other ways of measuring the visual information for IQA, rather than the signal-to-noise ratio. The information fidelity of the distorted image and the mutual information shared by the distorted image and the reference are two widely known examples. Both the response-wise and the information-wise thoughts offer solutions to IQA in the matter that the inner structure of HVS is not exploited, so the solutions are mostly with good intuitiveness and simplicity.

Thirdly, many of the known properties of HVS have been concluded as abstract concepts and specific research fields have been set up. For two very important instances, visual saliency (VS) and just noticeable difference (JND) have been introduced in Sect. 5.4. These effects of HVS can be classified as neither the response of certain parts or cells of HVS nor the response of the whole system. This is fine because the purpose of IQA is neither to study all details of HVS as thoroughly as possible nor to treat HVS as a complete Black box to only deem the interfaces as valuable. Reasonably, the abstract level should be moderate to avoid complex computations and to take care of HVS inner structures. Visual attention, including VS and JND, is a very good example of this moderate abstraction. VS or JND alone might not be a comprehensive description of the properties of HVS, but they are proved effective to be integrated with other feature extraction and pooling strategies to develop novel IQA methods. Moreover, because there are specific research fields of both VS and JND, many models are ready for us to employ, and the complication caused by studying lowest-level features is reduced. To conclude, it is promising to integrate our knowledge about HVS in neuroscience, psychology, and other related fields to build more accurate and robust IQA methods.

References

Achanta, R., Hemami, S., Estrada, F. & Susstrunk, S. (2009). Frequency-tuned salient region detection. In *Proceedings of IEEE Conference on Computer Vision and Pattern Recognition*, (pp. 1597–1604).

Ahumada, A. J., & Peterson, H. A. (1992). Luminance-model-based DCT quantization for color image compression. *Proc. SPIE on Human Vision, Visual Processing, and Digital Display III, 1666*, 365–374.

Avraham, T., & Lindenbaum, M. (2010). Esaliency (extended saliency): Meaningful attention using stochastic image modeling. *IEEE Transactions on Pattern Analysis and Machine Intelligence, 32*(4), 693–708.

Barten, P. (1999). *Contrast sensitivity of the human eye and its effects on image quality.* SPIE Press.

Baylor, D. A., Lamb, T. D., & Yau, K. W. (1979). Responses of retinal rods to single photons. *The Journal of Psychology, 288,* 613–634.

Bengio, Y. (2009). Learning deep hierarchies for AI. *Foundations and Trends in Machine Learning, 2*(1), 1–127.

Bex, P. J., & Makous, W. (2002). Spatial frequency, phase, and the contrast of natural images. *Journal of the Optical Society of America A, 19*(6), 1096–1106.

Bian, P., & Zhang, L. (2009). Biological plausibility of spectral domain approach for spatiotemporal visual saliency. *Advances in Neuro-Information Processing, 5506,* 251–258.

Blakemore, C., & Campbell, F. W. (1969). On the existence of neurones in the human visual system selectively sensitive to the orientation and size of retinal images. *The Journal of Physiology, 203*(1), 237–260.

Boring, E. G. (1942). *Sensation and perception in the history of experimental psychology.* New York: Appleton-Century.

Borji, A., & Itti, L. (2013). State-of-the-art in visual attention modeling. *IEEE Transactions on Pattern Analysis and Machine Intelligence, 35*(1), 185–207.

Brandao, T., & Queluz, M. P. (2008). No-reference image quality assessment based on DCT domain statistics. *Signal Processing, 88*(4), 822–833.

Bruce, N. D. B., & Tsotsos, J. K. (2009). Saliency, attention, and visual search: An information theoretic approach. *Journal of Vision, 9*(3), 1–24.

Burt, P. J. (1981). Fast filter transform for image processing. *Computer Graphics and Image Processing, 16*, 20–51.

Burt, P. J., & Adelson, E. H. (1983). The Laplacian pyramid as a compact image code. *IEEE Transactions on Communications, 9*(4), 532–540.

Chandler, D. M., & Hemami, S. S. (2007). VSNR: A wavelet-based visual signal-to-noise ratio for natural images. *IEEE Transactions on Image Processing, 16*(9), 2284–2298.

Chang, H., Zhang, Q., Wu, Q., & Gan, Y. (2015). Perceptual image quality assessment by independent feature detector. *Neurocomputing, 151*(3), 1142–1152.

Chiu, Y. J., & Berger, T. (1999). A software-only videocodec using pixelwise conditional differential replenishment and perceptual enhancements. *IEEE Transactions on Circuits and Systems for Video Technology, 9*(3), 438–450.

Chou, C. H., & Li, Y. C. (1995). A perceptually tuned subband image coder based on the measure of just-noticeable-distortion profile. *IEEE Transactions on Circuits on Systems for Video Technology, 5*(6), 467–476.

Cooley, J. W., & Tukey, J. W. (1965). An algorithm for the machine calculation of complex Fourier series. *Mathematics of Computation, 19*(90), 297–301.

Corporation, S. (1997). Sarnoff JND vision model. Contribution to IEEE G-2.1.6 Compression and Processing Subcommittee.

Crowley, J. L. (1981). *A representation for visual information* (Technique Report CMU-RI-TR-82-07). Pennsylvania: Robotics Institute, Carnegie-Mellon University.

Crowley, J. L., & Parker, A. C. (1984). A representation for shape based on peaks and ridges in the difference of low-pass transform. *IEEE Transactions on Pattern Recognition and Machine Intelligence, 6*(2), 156–170.

Crowley, J. L., & Sanderson, A. C. (1987). Multiple resolution representation and probabilistic matching of 2-D gray-scale shape. *IEEE Transactions on Pattern Recognition and Machine Intelligence, 9*(1), 113–121.

D'Angelo, A., Li, Z., & Barni, M. (2010). A full-reference quality metric for geometrically distorted images. *IEEE Transactions on Image Processing, 19*(4), 867–881.

Daly, S. (1992). The visible difference predictor: An algorithm for the assessment of image fidelity. *Proceedings of SPIE, 1616*, 2–15.

De Valois, R. L., Albrecht, D. G., & Thorell, L. G. (1982). Spatial frequency selectivity of cells in the macaque visual cortex. *Vision Research, 22*(5), 545–559.

Dickinson, S., Leonardis, A., Schiele, B., & Tarr, M. J. (2009). *Objective categorization: Computer and human vision perspectives.* Cambridge: Cambridge University Press.

Ding, Y., Wang, S., & Zhang, D. (2014a). Full-reference image quality assessment using statistical local correlation. *Electronics Letters, 50*(2), 79–80.

Ding, Y., Zhang, Y., Wang, X., Yan, X., & Krylov, A. S. (2014b). Perceptual image quality assessment metric using mutual information of Gabor features. *Science China: Information Science, 57*(3), 032111.

Ding, Y., Zhao, X., Zhang, Z., & Dai, H. (2017a). Image quality assessment based on multi-order local features description, modeling and quantification. *IEICE Transactions on Information and Systems, E, 100D*(6), 1303–1315.

Ding, Y., Zhao, Y., & Zhao, X. (2017b). Image quality assessment based on multi-feature extraction and synthesis with support vector regression. *Signal Processing: Image Communication, 54*, 81–92.

Du, S., Yan, Y., & Ma, Y. (2016). Blind image quality assessment with the histogram sequence of high-order local derivative patterns. *Digital Image Processing, 55*, 1–12.

Engelke, U., Kaprykowsky, H., Zepernick, H. J., & Ndjiki-Nya, P. (2011). Visual attention in quality assessment. *IEEE Signal Processing Magazine, 28*(6), 50–59.

Farias, M. C. Q., & Akamine, W. Y. L. (2012). On performance of image quality metrics enhanced with visual attention computational models. *Electronics Letters, 48*(11), 631–633.

Faugeras, O. D., & Pratt, W. K. (1980). Decorrelation methods of texture feature extraction. *IEEE Transactions on Pattern Analysis and Machine Intelligence, 2*(4), 323–332.

Felleman, D., & Essen, D. C. (1991). Distributed hierarchical processing in primate cerebral cortex. *Cerebral Cortex, 1,* 1–47.

Field, D. J. (1987). Relations between the statistics of natural images and the response properties of cortical cells. *Journal of Optical Society of America A, 4*(12), 2379–2397.

Freitas, P. G., Akamine, W. Y. L. & Farias, M. C. Q. (2016). No-reference image quality assessment based on statistics of local ternary pattern. In *8th International Conference on Quality of Multimedia Experience,* June 6–8, Lisbon, Portugal.

Gagalowicz, A. (1981). A new method for texture fields synthesis: Some applications to the study of human vision. *IEEE Transactions on Pattern Analysis and Machine Intelligence, 3*(5), 520–533.

Gao, X., Gao, F., Tao, D., & Li, X. (2013). Universal blind image quality assessment metrics via natural scene statistics and multiple kernel learning. *IEEE Transactions on Neural Networks and Learning Systems, 24*(12), 2013–2026.

Gao, X., Lu, W., Tao, D., & Li, X. (2009). Image quality assessment based on multiscale geometric analysis. *IEEE Transactions on Image Processing, 18*(7), 1409–1423.

Garcia-Diaz, A., Fdez-Vidal, X. R., Pardo, X. M., & Dosil, R. (2012). Saliency from hierarchical adaptation through decorrelation and variance normalization. *Image and Vision Computing, 30* (1), 51–64.

Gdyczynski, C. M., Manbachi, A., Hashemi, S., Lashkari, B., & Cobbold, R. S. C. (2014). On estimating the directionality distribution in pedicle trabecular bone from micro-CT images. *Physiological Measurement, 35*(12), 2415–2428.

Girod, B. (1993). What's wrong with mean-squared error? In *Visual factors of electronic image communications.* Cambridge: MIT Press.

Gu, K., Liu, M., Zhai, G., Yang, X., & Zhang, W. (2015). Quality assessment considering viewing distance and image resolution. *IEEE Transactions on Broadcasting, 61*(3), 520–531.

Guo, C., Ma, Q., & Zhang, L. (2008). Spatio-temporal saliency detection using phase spectrum of quaternion Fourier transform. In *Proceedings of IEEE Computer Society Conference on Computer Society Conference on Computer Vision and Pattern Recognition,* (pp. 1–8).

Guo, C., & Zhang, L. (2010). A novel multiresolution spatiotemporal saliency detection model and its applications in image and video compression. *IEEE Transactions on Image Processing, 19* (1), 185–198.

Hahn, P. J., & Mathews, V. J. (1998). An analytical model of the perceptual threshold function for multichannel image compression. *Proceedings of IEEE International Conference on Image Processing, 3,* 404–408.

Harel, J., Koch, C., & Perona, P. (2007). Graph-based visual saliency. In *Advances in Neural Information Processing Systems 19, Proceedings of the 2006 Conference,* (pp. 545–552).

Hecht, S., Shlar, S., & Pirenne, M. H. (1942). Energy, quanta, and vision. *Journal of General Physiology, 25,* 819–840.

Heeger, D., & Bergen, J. (1995). Pyramid-based texture analysis/synthesis. In *Proceeding of ACM SIGGRAPH,* (pp. 229–238).

Hong, R., Pan, J., Hao, S., Wang, M., Xue, F., & Wu, X. (2014). Image quality assessment based on matching pursuit. *Information Sciences, 273,* 196–211.

Hou, X., & Zhang, L. (2007). Saliency detection: A spectral residual approach. In *IEEE Conference on Computer Vision and Pattern Recognition,* (pp. 2280–2287).

Hubel, D. H., & Wiesel, T. N. (1962). Receptive fields, binocular interaction and functional architecture in the cat's visual cortex. *Journal of Physiology, 160*(1), 106–154.

Hubel, D. H., & Wiesel, T. N. (1968). Receptive fields and functional architecture of monkey striate cortex. *Journal of Physiology, 195*(1), 215–243.

Hurvich, L. (1981). *Color vision.* Sunderland: Sinauer Associates Inc.

Hyvärinen, A., Hurri, J., & Hoyer, P. O. (2009). *Natural image statistics: A probabilistic approach to early computational vision*. Berlin: Springer.

Itti, L., & Baldi, P. (2009). Bayesian surprise attracts human attention. *Vision Research, 49*(10), 1295–1306.

Itti, L., Koch, C., & Niebur, E. (1998). A model of saliency-based visual attention for rapid scene analysis. *IEEE Transactions on Pattern Analysis and Machine Intelligence, 20*(11), 1254–1259.

Jia, Y., Lin, W., & Kassim, A. A. (2006). Estimating just-noticeable distortion for video. *IEEE Transactions on Circuits and Systems for Video Technology, 16*(7), 820–829.

Jiang, Q., Shao, F., Jiang, G., Yu, M., & Peng, Z. (2015). Supervised dictionary learning for blind image quality assessment using quality-constraint sparse coding. *Journal of Visual Communication and Image Representation, 33*, 123–133.

Kandel, E. R., Schwartz, J. H., & Jessel, T. M. (2000). *Principles of neural sciences*. New York: McGraw-Hill.

Kingdom, F. A. A., Hayes, A., & Field, D. J. (1995). Sensitivity to contrast histogram differences in synthetic wavelet-textures. *Vision Research, 41*(5), 585–598.

Koch, C., & Ullman, S. (1985). Shifts in selective visual attention: Towards the underlying neural circuitry. *Human Neurobiology, 4*(4), 219–227.

Kremers, J. (2005). *The primate visual system: A comparative approach*. Hoboken: Wiley.

Krüger, N., Janssen, P., Kalkan, S., Lappe, M., Leonardis, A., Piater, J., et al. (2013). *IEEE Transactions on Pattern Analysis and Machine Intelligence, 35*(8), 1847–1871.

Larson, E. C., & Chandler, D. M. (2008). Unveiling relationships between regions of interest and image fidelity metrics. *Proceedings of the Society of Photo-optical Instrumentation Engineers, 6822*: 6822A1-16.

Larson, E. C., & Chandler, D. M. (2010). Most apparent distortion: Full-reference image quality assessment and the role of strategy. *Journal of Electronic Imaging, 19*(1), 011006.

Le Meur, O., Le Callet, P., Barba, D., & Thoreau, D. (2006). A coherent computational approach to model bottom-up visual attention. *IEEE Transactions on Pattern Analysis and Machine Intelligence, 28*(5), 802–817.

Legras, R., Chanteau, N., & Charman, W. N. (2004). Assessment of just-noticeable differences for refractive errors and spherical aberration using visual simulation. *Optometry and Vision Science, 81*(9), 718–728.

Lewis, J., & Essen, D. C. (2000). Corticocortical connections of visual, sensorimotor, and multimodal processing areas in the parietal lobe of the macaque monkey. *Journal of Comparative Neurology, 428*(1), 112–137.

Li, J., Duan, L. Y., Chen, X., Huang, T., & Tian, Y. (2015). Finding the secret of image saliency in the frequency domain. *IEEE Transactions on Pattern Analysis and Machine Intelligence, 37* (12), 2428–2440.

Li, Q., & Wang, Z. (2009). Reduced-reference image quality assessment using divisive normalization-based image representation. *IEEE Journal of Selected Topics in Signal Processing, 3*(2), 202–211.

Li, S., Zhang, F., Ma, L., & Ngan, K. N. (2011). Image quality assessment by separately evaluating detail losses and additive impairments. *IEEE Transactions on Multimedia, 13*(5), 935–949.

Li, X., Tao, D., Gao, X., & Lu, W. (2009). A natural image quality evaluation metric. *Signal Processing, 89*(4), 548–555.

Lin, J., Liu, T., Lin, W., & Kuo, C. (2013). Visual-saliency-enhanced image quality assessment indices. In *Asia-Pacific Signal and Information Processing Association Annual Summit and Conference*, (pp. 1–4).

Lin, W., Dong, L., & Xue, P. (2003). Discriminative analysis of pixel difference towards picture quality prediction. *Proceedings of IEEE International Conference on Image Processing, 3*, 193–196.

Liu, H., & Heynderickx, I. (2011). Visual attention in objective image quality assessment: Based on eye-tracking data. *IEEE Transactions on Circuits and Systems for Video Technology, 21*(7), 971–982.

Liu, L., Dong, H., Huang, H., & Bovik, A. C. (2014a). No-reference image quality assessment in curvelet domain. *Signal Processing: Image Communication, 29*(4), 494–505.

Liu, M., & Yang, X. (2009). Image quality assessment using contourlet transform. *Optical Engineering, 48*(10), 107201.

Liu, X., Sun, C., & Yang, L. T. (2015). DCT-based objective quality assessment metric of 2D/3D image. *Multimedia Tools and Applications, 74*(8), 2803–2820.

Liu, Z., Zou, W., & Le Meur, O. (2014b). Saliency tree: A novel saliency detection framework. *IEEE Transactions on Image Processing, 23*(5), 1937–1952.

Lu, W., Gao, X., Tao, D., & Li, X. (2008). A wavelet-based image quality assessment method. *International Journal of Wavelets Multiresolution and Information, 6*(4), 541–551.

Lu, W., Zeng, K., Tao, D., Yuan, Y., & Gao, X. (2010). No-reference image quality assessment in contourlet domain. *Neurocomputing, 73*(4–6), 784–794.

Lubin, J. (1993). The use of psychophysical data and models in the analysis of display system performance. In A. B. Watson (Ed.), *Digital Images and Human Vision* (pp. 163–178). Cambridge: MIT Press.

Lubin, J. (1995). A visual discrimination mode for image system design and evaluation. In E. Peli (Ed.), *Visual models for target detection and recognition* (pp. 207–220). Singapore: World Scientific Publishers.

Ma, L., Li, S., & Ngan, K. N. (2013). Reduced-reference image quality assessment in reorganized DCT domain. *Signal Processing: Image Communication, 28*(8), 884–902.

Ma, L., Li, S., Zhang, F., & Ngan, K. N. (2011). Reduced-reference image quality assessment using reorganized DCT-based image representation. *IEEE Transactions on Multimedia, 13*(4), 824–829.

Ma, L., Wang, X., Liu, Q., & Ngan, K. N. (2016). Reorganized DCT-based image representation for reduced reference stereoscopic image quality assessment. *Neurocomputing, 215*(SI), 21–31.

Ma, Q., & Zhang, L. (2008). Saliency-based image quality assessment criterion. *Advanced Intelligent Computing Theories and Applications, International Conference on Intelligent Computing, 5226*, 1124–1133.

Ma, Q., Zhang, L., & Wang, B. (2010). New strategy for image and video quality assessment. *Journal of Electronic Imaging, 19*(1), 1–14.

Manduchi, R., Perona, P., & Shy, D. (1997). Efficient deformable filter banks. *IEEE Transactions on Signal Processing, 46*(4), 1168–1173.

Mannos, J. L., & Sakrison, D. J. (1974). The effects of a visual fidelity criterion on the images. *IEEE Transactions on Information Theory, 20*(4), 525–536.

Marat, S., Phuoc, T. H., Granjon, L., Guyader, N., Pellerin, D., & Guerin-Dugue, A. (2009). Modeling spatio-temporal saliency to predict gaze direction for short videos. *International Journal of Computer Vision, 82*(3), 231–243.

Marr, D. (1977). Vision: A computational investigation into the human representation and processing of visual information. Freeman.

Maunsell, J. H. R., & Essen, D. C. (1983). Functional properties of neurons in middle temporal area of the macaque monkey, I, selectivity for stimulus direction, speed, and orientation. *Journal of Neurophysiology, 49*(5), 1127–1147.

McCann, J. J., McKee, S. P., & Taylor, T. H. (1976). Quantitative studies in retinex theory a comparison between theoretical predictions and observer responses to the "color mondrian" experiments. *Vision Research, 16*(5), 445–458.

Meer, P., Baugher, E. S., & Rosenfeld, A. (1987). Frequency domain analysis and synthesis of image generating kernels. *IEEE Transactions on Pattern Recognition and Machine Intelligence, 9*(4), 512–522.

Moorthy, A. K., & Bovik, A. C. (2009). Visual importance pooling for image quality assessment. *IEEE Journal of Selected Topics in Signal Processing, 3*(2), 193–201.

Moulden, B., Kingdom, F. A. A., & Gatley, L. F. (1990). The standard deviation of luminance as a metric for contrast in random-dot images. *Perception, 19*(1), 79–101.

Nakamura, H., Kuroda, T., Wakita, M., Kusunoki, M., Kato, A., Mikami, A., et al. (2001). From three-dimensional space vision to prehensile hand movements: The lateral intraparietal area links the area V3A and the anterior intraparietal area in macaques. *Journal of Neuroscience, 21* (20), 8174–8187.

Narwaria, M., Lin, W., McLoughlin, I. V., Emmanuel, S., & Chia, L. T. (2012). Fourier transform-based scalable image quality measure. *IEEE Transactions on Image Processing, 21* (8), 3364–3377.

Ojala, T., Pietikäinen, M., & Harwood, D. (1996). A comparative study of texture measures with classification based on feature distributions. *Pattern Recognition, 29*, 51–59.

Ojala, T., Pietikäinen, M., & Mäenpää, T. (2002). Multiresolution gray-scale and rotation invariant texture classification with local binary pattern. *IEEE Transactions on Pattern Analysis and Machine Intelligence, 24*(7), 971–987.

Orban, G. A. (2008). Higher order visual processing in macaque extrastriate cortex. *Physiological Reviews, 88*, 59–89.

Pati, P. B., & Ramakrishnan, A. G. (2008). Word level multi-script identification. *Pattern Recognition Letters, 29*, 1218–1229.

Peters, R., Iyer, A., Itti, L., & Koch, C. (2005). Components of bottom-up gaze allocation in natural images. *International Journal of Neural Systems, 45*(18), 2397–2416.

Portilla, J., & Simoncelli, E. P. (2000). A parametric texture model based on joint statistics of complex wavelet coefficients. *International Journal of Computer Vision, 40*, 49–71.

Poynton, C. (1998). The rehabilitation of gamma. In *Proceedings of SPIE Human Vision and Electronic Imaging*, pp. 232–249.

Prewitt, J. M. S. (1970). Object enhancement and extraction. In B. S. Lipkin & A. Rosenfeld (Eds.), *Picture processing and psychopictorics*. Cambridge: Academic Press.

Qian, J., Wu, D., Li, L., Cheng, D., & Wang, X. (2014). Image quality assessment based on multi-scale representation of structure. *Digital Signal Processing, 33*, 125–133.

Raju, S. S., Pati, P. B., & Ramakrishnan, A. G. (2004). Gabor filter based block energy analysis for text extraction from digital document images. In *Proceedings of the 1st International Workshop on Document Image Analysis for Libraries*, (pp. 233–243).

Ramasubramanian, M., Pattanaik, S. N., & Greenberg, D. P. (1999). A perceptually based physical error metric for realistic image synthesis. In *Proceedings of International Conference on Computer Graphics and Interactive Techniques*, (pp. 73–82).

Ramos, M. G., & Hemami, S. S. (2001). Suprathreshold wavelet coefficient quantization in complex stimuli: Psychophysical evaluation and analysis. *Journal of the Optical Society of America A, 18*(10), 2385–2397.

Rezazadeh, S., & Coulombe, S. (2013). A novel discrete wavelet transform framework for full reference image quality assessment. *Signal, Image and Video Processing, 7*(3), 559–573.

Roberts, L. G. (1965). Machine perception of three-dimensional solids. In J. T. Tippet (Ed.), *Optical and electro-optical information processing*. Cambridge: MIT Press.

Rodríguez-Sánchez, A. J., Simine, E., & Tsotsos, J. (2007). Attention and visual search. *International Journal of Neural Systems, 17*(4), 275–288.

Saad, M. A., Bovik, A. C., & Charrier, C. (2010). A DCT statistics-based blind image quality index. *IEEE Signal Processing Letters, 17*(6), 583–586.

Saad, M. A., Bovik, A. C., & Charrier, C. (2012). Blind image quality assessment: A natural scene statistics approach in the DCT domain. *IEEE Transactions on Image Processing, 21*(8), 3339–3352.

Saha, A., & Wu, Q. M. J. (2013). Perceptual image quality assessment using phase deviation sensitive energy features. *Signal Processing, 93*(11), 3182–3191.

Safranek, R. J., & Johnston, J. D. (1989). A perceptually tuned sub-band image coder with image dependence quantization and post-quantization data compression. In *Proceedings of IEEE Conference on Acoustic, Speech, and Signal Processing*, (pp. 1945–1948).

Sezan, M. I., Yip, K. L., & Daly, S. (1987). Uniform perceptual quantization: Applications to digital radiography. *IEEE Transactions on Systems, Man, and Cybernetics, 17*(4), 622–634.

Sheikh, H. R., & Bovik, A. C. (2006). Image information and visual quality. *IEEE Transactions on Image Processing, 15*(2), 430–444.

Sheikh, H. R., Bovik, A. C., & Cormack, L. (2005a). No-reference quality assessment using natural scene statistics: JPEG2000. *IEEE Transactions on Image Processing, 14*(11), 1918–1927.

Sheikh, H. R., Bovik, A. C., & de Vaciana, G. (2005b). An information fidelity criterion for image quality assessment using natural scene statistics. *IEEE Transactions on Image Processing, 14* (12), 2117–2128.

Sheikh, H. R., Sabir, M. F., & Bovik, A. C. (2006). A statistical evaluation of recent full reference image quality assessment algorithms. *IEEE Transactions on Image Processing, 15*(11), 3440–3451.

Shen, J., Li, Q., & Erlebacher, G. (2011). Hybrid no-reference natural image quality assessment of noisy, blurry, JPEG2000, and JPEG images. *IEEE Transactions on Image Processing, 20*(8), 2089–2098.

Sobel, I. E. (1970). Camera models and machine perception. *Ph.D. Dissertation*. California: Stanford University.

Soundararajan, R., & Bovik, A. C. (2012). RRED indices: Reduced reference entropic differencing for image quality assessment. *IEEE Transactions on Image Processing, 21*(2), 517–526.

Stromeyer, C. F., & Julesz, B. (1972). Spatial-frequency masking in vision: Critical bands and spread of masking. *Journal of the Optical Society of America, 62*(10), 1221–1232.

Tan, X., & Triggs, B. (2010). Enhanced local texture feature sets for face recognition under difficult lighting conditions. *IEEE Transactions on Image Processing, 19*(6), 1635–1650.

Tanaka, K. (1996). Inferotemporal cortex and object vision. *Annual Review of Neuroscience, 19* (1), 109–139.

Tanaka, K., & Saito, H. A. (1989). Analysis of motion of the visual field by direction, expansion/contraction, and rotation cells clustered in the dorsal part of the medial superior temporal area of the macaque monkey. *Journal of Neurophysiology, 62*(3), 626–641.

Tenenbaum, F. E., David, S. V., Singh, N. C., Hsu, A., Vinje, W. E., & Gallant, J. L. (2001). Estimating spatio-temporal receptive fields of auditory and visual neurons from their responses to natural stimuli. *Network: Computation in Neural Systems, 12*(3), 289–316.

Teo, P. C., & Heeger, D. J. (1994). Perceptual image distortion. *Proceedings of SPIE, 2179*, 127–141.

Tiippana, K., Näsänen, R., & Rovamo, J. (1994). Contrast matching of two-dimensional compound gratings. *Vision Research, 34*(9), 1157–1163.

Toet, A. (2011). Computational versus psychophysical bottom-up image saliency: A comparative evaluation study. *IEEE Transactions on Pattern Analysis and Machine Intelligence, 33*(11), 2131–2146.

Tong, Y., Konik, H., Cheikh, F. A., & Tremeau, A. (2010). Full reference image quality assessment based on saliency map analysis. *Journal of Imaging Science and Technology, 54* (3), 30503:1–30503:14.

Toprak, S., & Yalman, Y. (2017). A new full-reference image quality metric based on just noticeable difference. *Computer Standards & Interfaces, 50*, 18–25.

Torralba, A. (2003). Modeling global scene factors in attention. *Journal of the Optical Society of America A-Optics Image Science and Vision, 20*(7), 1407–1418.

Ungerleider, L. G., & Mishkin, M. (1982). Two cortical visual systems. In D. J. Ingle, M. A. Goodale, & R. J. W. Mansfield (Eds.), *Analysis of visual behavior* (pp. 549–586). Cambridge: MIT Press.

Uzair, M., & Dony, R. D. (2017). Estimating just-noticeable distortion for images/videos in pixel domain. *IET Image Processing, 11*(8), 559–567.

Van Nes, F. L., & Bouman, M. A. (1967). Spatial modulation transfer in the human eye. *Journal of the Optical Society of America, 57*(3), 401–406.

Walther, D., & Koch, C. (2006). Modeling attention to salient proto-objects. *Neural Networks, 19* (9), 1395–1407.

Wang, C., Shen, M., & Yao, C. (2015). No-reference quality assessment for DCT-based compressed image. *Journal of Visual Communication and Image Representation, 28,* 53–59.

Wang, S., Gu, K., Ma, S., Lin, W., Liu, X., & Gao, W. (2016). Guided image contrast enhancement based on retrieved images in cloud. *IEEE Transactions on Multimedia, 18*(2), 219–232.

Wang, Z., & Bovik, A. C. (2006). *Modern image quality assessment.* New York: Morgan & Claypool.

Wang, Z., & Bovik, A. C. (2009). Mean squared error: Love it or leave it? A new look at signal fidelity measures. *IEEE Signal Processing Magazine, 26*(1), 98–117.

Wang, Z., Bovik, A. C., Sheikh, H. R., & Simoncelli, E. P. (2004). Image quality assessment: From error visibility to structural similarity. *IEEE Transactions on Image Processing, 13*(4), 600–612.

Watson, A. B. (1987a). Estimation of local spatial scale. *Journal of the Optical Society of America A, 5*(4), 2401–2417.

Watson, A. B. (1987b). The Cortex transform: Rapid computation of simulated neural images. *Computer Vision Graphics and Image Processing, 39*(3), 311–327.

Watson, A. B., & Ahumanda, A. (2005). A standard model for foveal detection of spatial contrast. *Journal of Vision, 5*(9), 717–740.

Watson, A. B., Yang, G. Y., Solomon, J. A., & Villasenor, J. (1997). Visibility of wavelet quantization noise. *IEEE Transactions on Image Processing, 6*(8), 1164–1175.

Wei, Z., & Ngan, K. N. (2009). Spatio-temporal just noticeable distortion profile for grey scale image/video in DCT domain. *IEEE Transactions on Circuits and Systems for Video Technology, 19*(3), 337–346.

Wen, Y., Li, Y., Zhang, X., Shi, W., Wang, L., & Chen, J. (2017). A weighted full-reference image quality assessment based on visual saliency. *Journal of Visual Communication and Image Representation, 43,* 119–126.

Willmore, B. D. B., Prenger, R. J., & Gallant, J. L. (2010). Neural representation of natural images in visual area V2. *Journal of Neuroscience, 30*(6), 2102–2114.

Wilson, H., & Bergen, J. (1979). A four-mechanism model for threshold spatial vision. *Vision Research, 19*(1), 19–32.

Wu, H. R., & Rao, K. R. (2005). *Digital image video quality and perceptual coding.* Florida: CRC Press.

Wu, Q., Li, H., Meng, F., Ngan, K. N., Luo, B., Huang, C., et al. (2016). Blind image quality assessment based on multichannel feature fusion and label transfer. *IEEE Transactions on Circuits and Systems for Video Technology, 26*(3), 425–440.

Xue, W., Mou, X., Zhang, L., Bovik, A. C., & Feng, X. (2014). Blind image quality assessment using joint statistics of gradient magnitude and Laplacian features. *IEEE Transactions on Image Processing, 23*(11), 4850–4862.

Yang, X., Lin, W., Lu, Z., Ong, E. P., & Yao, S. (2003a). Just-noticeable-distortion profile with nonlinear additivity model for perceptual masking in color images. *Proceedings of IEEE International Conference on Acoustics, Speech, and Signal Processing, 3,* 609–612.

Yang, X., Lin, W., Lu, Z., Ong, E. P., & Yao, S. (2003b). Perceptually-adaptive hybrid video encoding based on just-noticeable-distortion profile. *Proceedings of the Society of Photo-optical Instrumentation Engineers, 5150,* 1448–1459.

Ye, P., & Doermann, D. (2012). No-reference image quality assessment using visual codebooks. *IEEE Transactions on Image Processing, 21*(7), 3129–3138.

Zhai, G., Zhang, W., Yang, X., Lin, W., & Xu, Y. (2008). No-reference noticeable blockiness estimation in images. *Signal Processing: Image Communication, 23*(6), 417–432.

Zhang, L., & Li, H. (2012). SR-SIM: A fast and high performance IQA index based on spectral residual. In *Proceedings of IEEE International Conference on Image Processing,* (pp. 1473–1476).

Zhang, L., Gu, Z., & Li, H. (2013a). SDSP: A novel saliency detection method by combining simple priors. In *Proceedings of IEEE International Conference on Image Processing*, (pp. 171–175).

Zhang, L., Shen, Y., & Li, H. (2014a). VSI: A visual saliency-induced index for perceptual image quality assessment. *IEEE Transactions on Image Processing, 23*(10), 4270–4281.

Zhang, L., Tong, M. H., Marks, T. M., Shan, H., & Cottrell, G. W. (2008a). SUN: A Bayesian framework for saliency using natural statistics. *Journal of Vision, 8*(7), 32.

Zhang, L., Zhang, L., Mou, X., & Zhang, D. (2011). FSIM: A feature similarity index for image quality assessment. *IEEE Transactions on Image Processing, 20*(8), 2378–2386.

Zhang, M., Mou, X., Fujita, H., Zhang, L., Zhang, X., & Xue, W. (2013b). Local binary pattern statistics feature for reduced reference image quality assessment. *Proceedings of SPIE, 8660* (86600L), 1–8.

Zhang, W., Borji, A., Wang, Z., Le Callet, P., & Liu, H. (2016). The application of visual saliency models in objective image quality assessment: A statistical evaluation. *IEEE Transactions on Neural Networks and Learning Systems, 27*(6), 1266–1278.

Zhang, X., Lin, W., & Xue, P. (2008b). Just-noticeable difference estimation with pixels in images. *Journal of Visual Communication and Image Representation, 19*(1), 30–41.

Zhang, Y., & Chandler, D. M. (2013). No-reference image quality assessment based on log-derivative statistics of natural scenes. *Journal of Electronic Imaging, 22*(4), 043025.

Zhang, Y., Moorthy, A. K., Chandler, D. M., & Bovik, A. C. (2014b). D-DIIVINE: No-reference image quality assessment based on local magnitude and phase statistics of natural scenes. *Signal Processing: Image Communication, 29*(7), 725–747.

Zhao, Y., Ding, Y., & Zhao, X. (2016). Image quality assessment based on complementary local feature extraction and quantification. *Electronics Letters, 52*(22), 1849–1850.

Zhu, W., Liang, S., Wei, Y., & Sun, J. (2014). Saliency optimization from robust background detection. In *Proceedings of IEEE Conference on Computer Vision and Pattern Recognition*, (pp. 2814–2821).

Chapter 6
Image Quality Assessment Based on Natural Image Statistics

6.1 Introduction

Since images are ultimately viewed by human beings, by taking advantage of the limited understanding of human visual system, it is meaningful to capture the features by reflecting the way that human beings perceive the image (Zhang et al. 2014). However, it is very difficult to model the complex and rigorous HVS well relying on the limited understanding upon it. In addition, the high computational complexity also limits the application of these methods.

In this scenario, most of the state-of-the-art IQA methods attempt to extract the statistical properties (features) of an image that are closely related to the image inherent quality (Zhao et al. 2016). Such methods are inclined to effectively extract quality-aware features and have achieved notable success (Sheikh et al. 2005; Moorthy and Bovik 2011; Saad et al. 2012). The most fundamental principle underlying natural image statistics-based IQA is that HVS is highly adapted to extract statistical information from the visual scene. And if there are some distortions introduced in an image, they may inevitably affect the statistical characteristics. In respect of image quality assessment, intuitively, the features' statistical characteristics of the distorted image are quite different from that of the reference image. Therefore, a measurement of the change of such statistical characteristics should provide a good approximation to perceptual image quality. Furthermore, the main motivation for exploring natural image statistics is the computational modeling of biological visual systems (Hyvärinen et al. 2009). A theoretical framework that considers the properties of the visual system to be reflections of the statistical structure of natural images because of evolutionary adaptation processes is gaining more and more support. It should be noted that *natural images* are photographs of the typical environment where we live.

Depending on how statistics information is modeled, there may be different ways to develop IQA algorithms. In this chapter, we first introduce the methods based on structural similarity which are based on the hypothesis that HVS is highly adapted

© Zhejiang University Press, Hangzhou and Springer-Verlag GmbH Germany 2018
Y. Ding, *Visual Quality Assessment for Natural and Medical Image*,
https://doi.org/10.1007/978-3-662-56497-4_6

for extracting structural information from images. Then, we introduce the methods that extract quality-aware features based on multifractal analysis. Subsequently, methods with local textural information extraction are given in Sect. 6.4. Finally, in Sect. 6.5, we discuss the usage of independent component analysis in IQA. It is really worthy to point out that exploiting the image information jointly in different domains can provide richer clues, which are not evident in any one individual domain (Oszust 2016). Therefore, multidomain feature extraction by considering joint occurrences of two or more features is necessary and constructive (Zhao et al. 2016; Wu et al. 2016).

6.2   IQA Based on Structural Similarity

There is an assumption that HVS is adapted to extract the information of structures (Wang et al. 2004). Structural information can represent objects' structures with the dependence among the pixels, especially the pixels in neighborhoods. The natural images which refer to the reference images in IQA are found to be highly structured. In the meanwhile, the structure of higher-quality image closely matches natural structure. On the contrary, a lower-quality image fails to have the similar structural information as the natural one. Consequently, measuring the structural similarity is a feasible approach to assess image quality. However, the definition of effective structural information related to the perceived distortion is a great challenge.

Multidimensional scaling analysis (Goodman and Pearson 1979) is one of the early attempts in structural similarity judgment toward the multiply-impaired television pictures. In (Eskicioglu and Fisher 1995), there is the evaluation of many quality measures including "structural content." In order to evaluate the quality of compressed images, a metric named block-wise distortion measure was proposed in (Fränti 1998), which consists of three factors, structural errors, contrast errors, and quantization errors. In addition, the structural errors are based on the edge information. A universal image quality index is proposed to measure the image quality and is applied to various applications in image processing (Wang and Bovik 2002). This model is the integration of contrast distortion, luminance distortion, and loss of correlation. A measurement named structural similarity (SSIM) proposed in (Wang et al. 2004) is one of the most popular methods with numerous variations in IQA. The method is based on the universal image quality index.

In SSIM, the basic principle is the definition of structural and nonstructural distortions. Structural distortion can have influence on the structural information of the objects in images. On the contrary, nonstructural distortion cannot modify the objects' structure. As shown in Fig. 6.1, (a) is the reference image, and the distorted images include contrast stretched image (b), mean shifted image (c), blurred image (d), and image corrupted by white noise (e), with the same MSE. It is obvious that (b) and (c) are of higher quality comparing with (d) and (e) in visual perception. Consequently, the distortions in contrast and luminance are nonstructural. In this

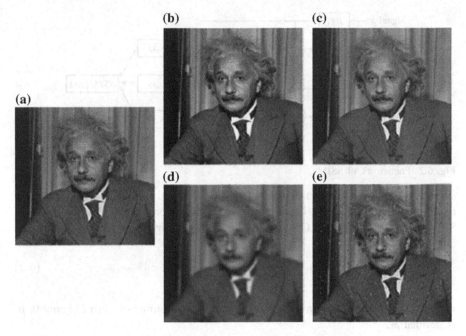

Fig. 6.1 Comparisons of structural distortions and nonstructural distortions

case, contrast and luminance are independent to structural information, which can be easily understood. When we are supposed to observe an object in the scene with different light conditions, the perceived images have different luminance and contrast, but the structure of the object remains unchanged. Therefore, SSIM algorithm applies three factors, luminance errors, contrast errors, and structure errors, on image quality assessment system and structural information modeled in SSIM is independent of luminance and contrast.

6.2.1 Structural Similarity

In SSIM (Wang et al. 2004), in order to measure the structural similarity between signal x and signal y, there are three comparisons including luminance comparison, contrast comparison, and structure comparison which are denoted as $l(x, y)$, $c(x, y)$, $s(x, y)$, respectively,

$$\text{SSIM}(x,y) = [l(x,y)]^{\alpha}[c(x,y)]^{\beta}[s(x,y)]^{\gamma} \tag{6.1}$$

where α, β, and γ are greater than 0. And the three parameters are employed to determine the relative importance of every component. In IQA, assume one of the input signals being reference image. Then, the quantitative measurement of the

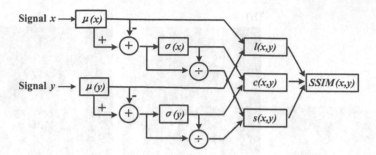

Fig. 6.2 Framework of SSIM

structural similarity between two signals can be obtained from (6.1). In addition, the result is expected to be consistent with subject assessment result. The framework of SSIM is shown in Fig. 6.2. Considering luminance comparison, $l(x, y)$ is defined as,

$$l(x, y) = \frac{2\mu_x\mu_y + C_1}{\mu_x^2 + \mu_y^2 + C_1} \tag{6.2}$$

where C_1 is a constant of small value to avoid the denominator in (6.2) being 0. μ_x is defined as,

$$\mu_x = \frac{1}{N}\sum_{i=1}^{N} x_i \tag{6.3}$$

where N is the total number of the primaries in discrete signal x and x_i denotes one element. Actually, μ_x is the mean intensity of signal x. And, the definition of μ_y is the same as μ_x. Considering contrast comparison, mean intensity is removed from signal x, and the signal contrast σ_x can be calculated as,

$$\sigma_x = \left(\frac{1}{N-1}\sum_{i=1}^{N}(x_i - \mu_x)^2\right)^{1/2} \tag{6.4}$$

where the parameters are determined as above. The definition of signal contrast σ_y is in a similar form. Then, use σ_x and σ_y to construct contrast comparison

$$c(x, y) = \frac{2\sigma_x\sigma_y + C_2}{\sigma_x^2 + \sigma_y^2 + C_2} \tag{6.5}$$

where C_2 is a constant and $C_2 \ll 1$. The principle of structural comparison is that it is independent of another two comparisons. In SSIM, Wang et al. employ $(x - \mu_x)/\sigma_x$ and $(y - \mu_y)/\sigma_y$ to have zero-mean intensity and unit standard deviation. After luminance subtraction processing and variance normalization processing, SSIM metric treats the information preserved as the structural information. The formal definition of structural comparison is given as,

$$s(x, y) = \frac{\sigma_{xy} + C_3}{\sigma_x \sigma_y + C_3} \tag{6.6}$$

where C_3 is a small constant. And σ_{xy} is represented as,

$$\sigma_{xy} = \frac{1}{N-1} \sum_{i=1}^{N} (x_i - \mu_x)(y_i - \mu_y) \tag{6.7}$$

where the parameters are determined as above. In this case, SSIM consists of three comparisons as shown in (6.1). There is the simple expression in (Wang et al. 2004). The three important factors in (6.1) are set to 1, and C_3 in (6.6) is equal to $C_2/2$. In this case, there is the SSIM index in specific form,

$$\text{SSIM}(x, y) = \frac{(2\mu_x \mu_y + C_1)(2\sigma_{xy} + C_2)}{(\mu_x^2 + \mu_y^2 + C_1)(\sigma_x^2 + \sigma_y^2 + C_2)}. \tag{6.8}$$

There are three properties in SSIM (Wang et al. 2004). Firstly, since SSIM(x, y) = SSIM(y, x), the metric is symmetrical. Then, the result of SSIM cannot be greater than 1. Finally, this metric has unique maximum when x is equal to y ($x_i = y_i$, $i = 1$, $2,...,N$). These properties make it convenient to carry out quantitative measurement in IQA.

In IQA, the SSIM index prefers to be applied locally. This is based on the images' statistical properties and HVS properties. On the one hand, statistics on images globally neglects the variances of luminance and contrast in different local regions. On the other hand, because of the properties of foveated vision, human observer perceives only a local area in high resolution.

Consequently, the reasonable way to obtain the statistical features is in local regions and the image is divided into patches by a local window moving pixel by pixel. Actually, μ_x, σ_x, μ_y, σ_y, and σ_{xy} are supposed to be calculated in local window. However, using windows leads to an undesirable problem called "blocking" artifacts (Wang et al. 2004), and the solution is to apply the circular-symmetric Gaussian weighting function ($\sum Ni = 1\omega_i = 1$) on the primaries in local region, taking μ_x as example,

$$\mu_x = \sum_{i=1}^{N} \omega_i x_i \tag{6.9}$$

which has extensive use in (Wang 2001; Wang and Bovik 2002). After the windowing approach, SSIM index becomes the similarity measurement of local regions. These indexes are combined to obtain the overall quality of image. Mean SSIM (MSSIM) is used to obtain the final result,

$$\mathrm{MSSIM}(X, Y) = \frac{1}{M} \sum_{j=1}^{M} \mathrm{SSIM}(x_j, y_j) \tag{6.10}$$

where X and Y are reference and distorted images, respectively. M is the total number of patches in image, and x_j refers to the contents in jth patch in reference image which is corresponding to the definition of y_j. Using the average value of the SSIM indexes of overall images is the basic method. The different weighting functions can be introduced into the SSIM method in different application (Privitera and Stark 2000; Rajashekar et al. 2003).

The display resolution has a close relation to viewing distance, sampling density, and the observer's perceptual capability. With these critical factors changing, there are many scales of resolutions. Therefore, the observed resolution has impact on evaluation of the image quality. SSIM is a single-scale model. In order to be adaptive to various resolution conditions, multiscale structural similarity (MS-SSIM) is proposed in (Wang et al., 2003). In this method, MS-SSIM regards the original image as scale 1, and the highest scale is denoted as S. The author employs MS-SSIM to measure the structural similarity between signal x and y,

$$\mathrm{MS} - \mathrm{SSIM}(x, y) = \prod_{j=1}^{S} \left[\mathrm{SSIM}_j(x, y) \right]^{\tau_j} \tag{6.11}$$

where j refers to jth scale and τ_j is the parameter to adjust the importance of each scale. In practice, the contrast comparison in (6.5) and structural comparison in (6.6), which are denoted as $c_j(x, y)$ and $s_j(x, y)$, respectively, are calculated in each scale and the difference is the luminance comparison only calculated at Mth scale,

$$\mathrm{MS} - \mathrm{SSIM}(x, y) = [l_S(x, y)]^{\alpha_S} \cdot \prod_{j=1}^{S} \left[c_j(x, y) \right]^{\beta_j} \left[s_j(x, y) \right]^{\gamma_j} \tag{6.12}$$

where β_j and γ_j are parameters of $c_j(x, y)$ and $s_j(x, y)$ at jth scale. $L_S(x, y)$ and α_S are only calculated at Sth scale. The determination of parameters refers to (Wang et al., 2003). It is proved that introducing multiscale into SSIM can construct more accurate measurement of image quality and the best performance is achieved when $S = 2$.

The early image quality assessment metrics focused on directly calculating the error between reference and distorted images. Then, many literatures found that it is important to construct IQA with the knowledge of HVS. In the meanwhile, metrics are proposed to simulate the function and properties of early HVS. HVS is so complex that it is very difficult to construct a complete model. And in order to have sufficient simulation, the features are extracted with many functions, which leads to the increasing of computation complexity.

SSIM is based on the assumption that HVS is adapted to extract the structural information. And in this case, the similarity of structure acts as the judgment of image quality. With the simple formulation, the application of it is tractable in image coding and optimization system. Experimental results show that the SSIM index has a better performance than early IQA metrics based on simulating HVS. SSIM is a complementary and alternative metric in IQA with statistical features applied. However, it is the initial attempt to relate structural similarity with image quality. There are many improvements and subsequent research about SSIM.

6.2.2 Complex Wavelet SSIM

It is a drawback of SSIM that the method is high sensitivity to geometrical distortions, including translation, rotation, scaling, which are not desired to be detected by IQA methods because the structure is not infected. However, typically caused by device movement during the acquisition of images, the geometrical distortions lead to the loss of the precise matching between compared images. The distortions listed above are nonstructural distortions, according to the definition of two categories of distortions mentioned above (see Sect. 6.2.1). In addition, a good IQA system should be able to remove the influence of these distortions, which seems difficult to SSIM index. Therefore, a complex wavelet SSIM (CW-SSIM) index (Sampat et al. 2009) is proposed as an extension to the SSIM metric. The original motivation of CW-SSIM is to construct an IQA metric that does not sensitive to these nonstructural geometrical distortions, based on the principle that phase would not be influenced by small geometric distortions. In this case, the properties of complex wavelet contain: (1) The magnitude and phase differences are separated; (2) it is more sensitive to phase difference than to magnitude difference; and (3) it loses the sensitivity to the relative phase difference.

There is a "mother wavelets" of symmetric complex wavelets that can be denoted as the modulation of low-pass filter,

$$w(u) = g(u)e^{jw_c u} \tag{6.13}$$

where w_c is the central frequency, and $g(u)$ is a slowly varying symmetric function. The family of "mother wavelets" are the versions with dilating, contracting, and translating, and they are written as,

$$w_{s,p}(u) = \frac{1}{\sqrt{s}}w\left(\frac{u-p}{s}\right) = \frac{1}{\sqrt{s}}g\left(\frac{u-p}{s}\right)e^{jw_c(u-p)/s} \tag{6.14}$$

where s denotes scale and p is translation factor. The continuous wavelet transform of $x(u)$ is described as,

$$X(s,p) = \frac{1}{2\pi}\int_{-\infty}^{\infty} X(w)\sqrt{s}G(sw - w_c)e^{jwp}dw \tag{6.15}$$

where $X(w)$ and $G(w)$ denote Fourier transforms of $x(u)$ and $g(u)$, respectively. Through sampling the continuous wavelet transform, discrete wavelet coefficients are obtained. Then employing $c_x = \{c_{x,i}|i = 1,\ldots,N\}$ and $c_y = \{c_{y,i}|i = 1,\ldots,N\}$ to represent the coefficients extracted from the two images in the same spatial position and the same wavelet sub-bands, respectively. The CW-SSIM index is defined in the similar form as SSIM index

$$\mathrm{CWSSIM}(c_x, c_y) = \frac{2\left|\sum_{i=1}^{N} c_{x,i}c_{y,i}^*\right| + C_4}{\sum_{i=1}^{N} |c_{x,i}|^2 + \sum_{i=1}^{N} |c_{y,i}|^2 + C_4} \tag{6.16}$$

where c^* is complex conjugate of c and C_4 is a constant of small value. In fact, there are two components in CW-SSIM index, where the first component represents magnitude distortions and the second reflects the consistent phase distortions.

$$\mathrm{CWSSIM}(c_x, c_y) = \frac{2\sum_{i=1}^{N} |c_{x,i}||c_{y,i}| + C_4}{\sum_{i=1}^{N} |c_{x,i}|^2 + \sum_{i=1}^{N} |c_{y,i}|^2 + C_4} \cdot \frac{2\left|\sum_{i=1}^{N} c_{x,i}c_{y,i}^*\right| + C_4}{2\sum_{i=1}^{N} |c_{x,i}||c_{y,i}| + C_4}.$$
$$\tag{6.17}$$

The magnitude of discrete wavelet coefficients determines the first component. And when the equation $|c_{x,j}| = |c_{y,j}|$ holds true for all i, the result reaches the maximum value 1. The second component is regarded as the key factor to detect the consistency of phase shifts between the two sets of coefficients c_x and c_y. And if the difference between $c_{x,i}$ and $c_{y,i}$ (for every i) is a constant, the second component reaches maximum value 1. Therefore, this component realizes the goal that CW-SSIM is insensitive to nonstructural distortions.

The proposer of the method proves that there are many distortions that CW-SSIM index insensitive to, including luminance distortion, contrast distortion, and geometric distortions as mentioned above. In fact, these distortions are all belonging to the nonstructural distortion. In this case, CW-SSIM addresses the SSIM's drawback that it is sensitive to geometric distortions. With a good performance on similarity measurement, CW-SSIM has a variety of applications in other fields, such as (1) comparing the result of image segmentation; (2) image registration; (3) object detection and tracking in video.

6.2.3   Information Content Weighted SSIM

Since luminance and contrast are space-variant in images, SSIM has a better performance in local feature extraction than in global feature extraction. Information content weighted SSIM (IW-SSIM) proposed in (Wang and Li 2011) takes advantage of the weights of local regions in multiscale SSIM. According to the knowledge about properties of HVS, observers cannot pay their attentions equally on each local region. The initial studies (Engelke et al. 2008; Larson and Chandler 2008; Larson et al. 2008) are focused on constructing models correlated with visual ROI (region-of-interest) detection and visual fixation. In practice, the subjective data of human fixation is difficult and inflexible to be acquired. Therefore, the authors are committed to use the statistical tendency of natural images. Moreover, they are focused on constructing a reliable computational model to quantitatively calculate the weights of local regions with the theoretical principles. There is an important hypothesis that HVS can be regarded as the optimal information extractor in computational vision science (Simoncelli and Olshausen 2001). Visual attention is correlated with the complexity of information content (Najemnik and Geisler 2005). And IW-SSIM adopts Gaussian scale mixture (GSM) to achieve the quantitative measurement.

In order to reduce the complexity of calculation and the high dimensionality of images, there is a Markov assumption that probability density of pixels (or coefficients of transform) is determined by the neighboring pixels (or coefficients). Since the neighborhood is composed of neighboring transform coefficients, GSM is proved (Wainwright and Simoncelli 1999) to be a good modeling for Markov assumption.

GSM is employed to work on the transform coefficients (such as, the coefficients of Laplacian pyramid transform or wavelet). Firstly, suppose $\mathbf{r}$ to be a column vector with a length of K,

$$\mathbf{r} = s\mathbf{u} \tag{6.18}$$

where $\mathbf{u}$ is Gaussian vector with zero mean, and s is mixing multiplier. In order to simplify the computation, Wang and Li set s as a fixed value instead of a random variable. Notice that the value of s is fixed at each local region and is varied from space and scale. Let $\mathbf{C}_{\mathbf{r}|s}$ and $\mathbf{C}_{\mathbf{u}}$ denote covariance matrix of $\mathbf{r}$ (with s fixed) and $\mathbf{u}$,

$$\mathbf{C}_{\mathbf{r}|s} = s^2\mathbf{C}_{\mathbf{u}} \tag{6.19}$$

Another fundamental theory of IW-SSIM is mutual information between the images. Mutual information can provide a more useful measurement than only extracting the information contained in a single image, which is corresponding to perceived signal in HVS. Assume $\mathbf{r}$ as the reference signal and $\mathbf{d}$ as the simplified distortion signal with gain factor g and Gaussian noise contamination $\mathbf{v}$,

$$\mathbf{d} = g\mathbf{r} + \mathbf{v} = gs\mathbf{u} + \mathbf{v} \tag{6.20}$$

where covariance $\mathbf{C}_{\mathbf{v}} = \sigma_v^2 \mathbf{I}$ and $\mathbf{I}$ are identity matrices. And the next step is processing the reference and distortion signals with the perceptual visual noise channel,

$$\mathbf{e} = \mathbf{r} + \mathbf{n}_1 = s\mathbf{u} + \mathbf{n}_1 \tag{6.21}$$

$$\mathbf{f} = \mathbf{d} + \mathbf{n_2} = gs\mathbf{u} + \mathbf{v} + \mathbf{n_2} \tag{6.22}$$

where $\mathbf{n_1}$ and $\mathbf{n_2}$ are independent white Gaussian noise with $\mathbf{C_{n1}} = \mathbf{C_{n2}} = \sigma_n^2\mathbf{I}$
According to (6.19), the $\mathbf{C_d}$, $\mathbf{C_e}$, and $\mathbf{C_f}$ can be described as,

$$\mathbf{C_d} = g^2 s^2 \mathbf{C_u} + \sigma_v^2 \mathbf{I} \tag{6.23}$$

$$\mathbf{C_e} = s^2 \mathbf{C_u} + \sigma_n^2 \mathbf{I} \tag{6.24}$$

$$\mathbf{C_f} = g^2 s^2 \mathbf{C_u} + \sigma_v^2 \mathbf{I} + \sigma_n^2 \mathbf{I}. \tag{6.25}$$

Here, Wang and Li aimed at calculating the total information perceived from reference and distortion images that is computed by the sum of mutual information $M(\mathbf{r}; \mathbf{e})$ and $M(\mathbf{d}; \mathbf{f})$, following the research in (Sheikh and Bovik 2006). And then, remove the common information denoted as $M(\mathbf{e}; \mathbf{f})$. Finally, the weight w is calculated as the result of total information content,

$$w = M(\mathbf{r}; \mathbf{e}) + M(\mathbf{d}; \mathbf{f}) - M(\mathbf{e}; \mathbf{f}). \tag{6.26}$$

According to the deducing process in (Wang and Li 2011), the information content weight w at one location and a specific scale can be presented as,

$$w = \frac{1}{2}\sum_{k=1}^{K} \log_2\left\{1 + \frac{\sigma_v^2}{\sigma_n^2} + \left(\frac{\sigma_v^2}{\sigma_n^4} + \frac{1+g^2}{\sigma_n^2}\right)s^2\lambda_k\right\} \tag{6.27}$$

where K is the length of vector and λ_k is the eigenvalues of $\mathbf{C_u}$, and $w_{i,j}$ represents the ith spatial location at jth scale. With the definition of MS-SSIM in (6.12), the contrast comparison and structural comparison are calculated all over scales, but luminance comparison only computed at M-scale. Therefore, overall IW-SSIM is calculated in the similar form as (6.11),

$$\text{IWSSIM} = \prod_{j=1}^{M} \left(IWSSIM_j\right)^{\tau_j} \tag{6.28}$$

where $M = 5$, $\{\tau_1, \tau_2, \tau_3, \tau_4, \tau_5\} = \{0.0448, 0.2856, 0.3001, 0.2363, 0.1333\}$ in (Wang et al. 2003), and

$$\text{IWSSIM}_j = \begin{cases} \dfrac{\sum_{i=1}^{N} w_{j,i} c(x_{j,i}, y_{j,i}) s(x_{j,i}, y_{j,i})}{\sum_{i=1}^{N} w_{j,i}} & j = 1, \ldots, M-1 \\ \dfrac{1}{N_j}\sum_{i=1}^{N} l(x_{j,i}, y_{j,i}) c(x_{j,i}, y_{j,i}) s(x_{j,i}, y_{j,i}) & j = M \end{cases}. \tag{6.29}$$

6.2.4 Feature Similarity

The application of SSIM in IQA is a good success, and it improves feature extraction methods from pixel-based ones to structure-based ones, which proves that the information perceived with HVS is structural. Some early relative studies (Marr and Hildreth 1980; Morrone and Burr 1988) show that although the information in images is complex, there are many redundant features, and the low-level feature is of great significance in perceived information. Therefore, a good IQA method can be designed to extract low-level features in reference and distortion images and to employ the difference between features extracted from reference and distortion images as the quantitative measurement. Zhang et al. proposed a low-level feature similarity (FSIM) metric for image quality assessment (Zhang et al. 2011).

It is proved in (Morrone and Burr 1988; Kovesi 1999; Henriksson et al. 2009) that phase congruency (PC) can be considered as a highly informative low-level feature, which is based on the assumption that the features are perceived at specific points, where the Fourier component is maximal in phase (Zhang et al. 2011). Since PC is contrast invariant, image gradient magnitude (GM) is employed as the complementary feature. FSIM is mainly constituted from the two features and achieves a good performance.

The definition of PC can be found in (Morrone et al. 1986), and the widely spread method used in FSIM is developed by (Kovesi 1999). Firstly, consider the 1D signal $g(x)$, and there are even symmetric and odd symmetric filters on scale j, denoted by E_j and O_j that constitute quadrature pair. The response vector of each quadrature pair at position x in scale j is described as,

$$[e_j(x), o_j(x)] = [g(x) * E_j, g(x) * O_j] \tag{6.30}$$

And the local amplitude in scale j denoted as A_j is calculated as,

$$A_j = \sqrt{e_j(x)^2 + o_j(x)^2} \tag{6.31}$$

$$M(x) = \sqrt{\left(\sum_j e_j(x)\right)^2 + \left(\sum_j o_j(x)\right)^2} \tag{6.32}$$

$$PC(x) = \frac{M(x)}{C_5 + \sum_j A_j(x)} \tag{6.33}$$

where C_5 is a small positive constant. Then, the log Gabor filters are employed as the filters mentioned above (i.e., E_j and O_j), and the transfer function in frequency domain is described as,

$$G(w) = \exp(-(\log(w/w_0)))^2 / 2\sigma_r^2) \tag{6.34}$$

where w is central frequency and σ_r are correlative to bandwidth of filter. Then, the analysis of 1D signal $g(x)$ can be extended to 2D grayscale images by combining the results from several directions with some spreading function. Gaussian is a widely used spread function and has many advantages (Kovesi 1999; Mancas-Thillou and Gosselin 2006; Wang et al. 2008). Next, applying Gaussian into (6.34), 2D log Gabor is shown as,

$$G_2(w, \theta_d) = \exp\left(-\frac{\left(\log\left(\frac{w}{w_0}\right)\right)^2}{2\sigma_r^2}\right) \exp\left(-\frac{(\theta - \theta_d)^2}{2\sigma_\theta^2}\right) \qquad (6.35)$$

where $\theta_d = d\pi/D$, D is the total number of directions, and $d = \{0,1,\ldots,D-1\}$ is the angle of filter; σ_θ controls the angular bandwidth of filter. Then, 2D PC with log Gabor is defined as,

$$PC_{2D}(x) = \frac{\sum_d M_{\theta_d}(x)}{C_6 + \sum_j \sum_d A_{j,\theta_d}(x)} \qquad (6.36)$$

where C_6 is a small positive constant, and

$$M_{\theta_d}(x) = \sqrt{\left(\sum_j e_{j,\theta_d}(x)\right)^2 + \left(\sum_j o_{j,\theta_d}(x)\right)^2} \qquad (6.37)$$

$$A_{j,\theta_d}(x) = \sqrt{\left(e_{j,\theta_d}(x)\right)^2 + \left(o_{j,\theta_d}(x)\right)^2} \qquad (6.38)$$

Gradient magnitude (GM) is defined as,

$$G = \sqrt{G_x^2 + G_y^2} \qquad (6.39)$$

where G_x and G_y are partial derivatives that can be calculated with Sobel operator (Jähne et al. 1999), Prewitt operator (Jain et al. 1995), and Scharr operator (Jähne et al. 1999).

For FSIM index calculation, the similarity between the reference image X and distorted image Y in each feature is measured. $S_{PC}(x, y)$ and $S_G(x, y)$ are the similarity measurements of PC and GM at the same location, respectively, described as,

$$S_{PC}(x,y) = \frac{2PC(x) \cdot PC(y) + C_7}{PC^2(x) + PC^2(y) + C_7} \qquad (6.40)$$

$$S_G(x,y) = \frac{2G(x) \cdot G(y) + C_8}{G^2(x) + G^2(y) + C_8}. \qquad (6.41)$$

where C_7 and C_8 are small positive constants. Thus, FSIM is defined as,

$$\text{FSIM} = \frac{\sum\limits_{i=1}^{N} S_{\text{PC},i}^2(x,y) \cdot S_{G,i}^2(x,y) \cdot \text{PC}_{\text{max},i}(x,y)}{\sum\limits_{i=1}^{N} \text{PC}_{\text{max},i}(x,y)} \tag{6.42}$$

where i refers to the ith location.

6.2.5   Summary

SSIM is based on the assumption that HVS is adapted to extract the information of structures (Wang et al. 2004) and the metric is composed of three complementary and independent components including luminance comparison, contrast comparison, and structure comparison. Since the structural information represents the objects' structure in images, SSIM index successfully introduces structure-based comparison (instead of pixel-based comparison) into IQA. With many variants achieving good performance, it can be regarded as a turning point in image quality assessment.

However, the initial version of SSIM index has some problems in application. Firstly, the local regions, where SSIM index calculates the difference between reference and distorted images, are given equal weights. There are many solutions to the problem, such as quality-based pooling (Moorthy and Bovik 2009), information content (Wang and Li 2011), and saliency mapping (Mendi 2015). Then, the estimations are not consistent among distortions of different types (such as blurred and noisy images) (Li and Bovik 2009). The tendency is to employ low-level feature similarity as quantitated measurement to IQA in the form similar to SSIM (Li and Bovik 2009; Zhang et al. 2011). Finally, SSIM with two images compared is applied in FR-IQA, which seldom happens in real world. In this case, the study concentrates on applying SSIM into RR-IQA in (Huang et al. 2011). In fact, the mainstream is NR-IQA. And there are many state-of-the-art NR-IQA metrics developed (Zhang and Roysam 2016; Gu et al. 2017; Zhou et al. 2017). Therefore, in order to construct an outstanding IQA metric, structural information in images is not sufficient. Other widely used approaches are introduced in the following sections.

6.3   IQA Based on Local Pattern Statistics

6.3.1   Local Pattern Statistics

Texture analysis has a wide range of applications in image processing, including segmentation (Yuan et al. 2015), classification (Ojala et al. 1996), and recognition

(Ahonen et al. 2006). As texture information can give the spatial arrangement of intensities, IQA can use the differences of texture information between reference and distorted images to measure image quality. Since texture information is mainly produced in primary visual cortex (V1 and V2) (Kruger et al. 2013), it is one of the low-level features that are more consistent with visual perception. The metrics based on texture information are believed to achieve a good performance.

Methods of texture extraction can be divided into many categories. Local pattern, observing the local structural texture information, is one of the research hot spots of texture analysis, which has the invariance of rotation and scales, low computational complexity, and low feature dimensionality. Local pattern statistics is based on the assumption that the complex texture is constituted of texture units and there are different distribution laws corresponding to different types of complex textures. And when it comes to the image quality assessment, the distribution of texture units is employed as features extracted from images. Since the texture of image can be altered by distortions, the differences between the reference and distorted images' distributions can reflect the degree of images being polluted.

Local binary pattern (LBP) is one of the typical local pattern statistics models. LBP uses the 0 pattern and 1 pattern to code the texture units, which is relative to the differences between the pixels and their neighborhoods. In this case, LBP is considered to retain only sign component of information in neighborhoods. However, it is proved that the importance of sign component is much greater than magnitude component (Guo et al. 2010) and the information extracted by LBP is corresponding to the local structural texture, such as edges, corners, and spots (Zhang et al. 2013), which are relative to visual perception. With the local structural information extracted, the natural images are supposed to follow a standard distribution that would be altered by distortions, which is fundamental to quantitative measurement. With the development of computer vision and recognition, many variants of LBP are proposed to complete LBP and extend the applications of LBP. For example, completed local binary pattern (CLBP) is proposed to combine the magnitude local binary pattern and center local binary pattern with local binary pattern to extract more complete texture information (Guo et al. 2010); local ternary pattern (LTP) uses ternary pattern instead of binary pattern to reduce the influence of imperceptible variation (Tan and Triggs 2010). For its high efficiency and low computational complexity, LBP and its variants have a wide range of applications (Ahonen et al. 2006; Tan and Triggs 2010; Luo et al. 2016). Then, we are supposed to provide an introduction of LBP and its excellent variants, as well as their application in IQA, to enlighten the researchers in IQA.

6.3.2 Local Binary Pattern

There are many problems in texture analysis in real world, such as invariance of spatial scale, invariance of orientation, invariance of gray scale. Local binary pattern (LBP) with uniform pattern and rotation invariance pattern is proposed to adapt to the complex application sceneries of texture analysis (Ojala et al. 2002) and has been successfully used in face recognition in (Ahonen et al. 2006). The foundation of LBP is the conception of texture spectrum proposed in (Wang and He 1990). Since the complex texture is composed of texture units, the distribution of these units can be obtained as the texture spectrum of the overall image. In texture analysis, the texture spectrum is successfully used to classify the textures. For image quality assessment, when the image is distorted, the texture information would be changed as well, so does the texture spectrum. Therefore, metrics that extract texture spectrum from images to assess image quality are appeared. However, the initial texture units in texture spectrum (Wang and He 1990) are three-level patterns. LBP is proposed to simplify the three-level patterns, and it reduces the amount of texture units from 3^8 to 2^8 in a 3×3 neighborhood. So local binary pattern attracts increasing attention in computer vision (Ojala et al. 1996). The coding of micro-pattern in a 3×3 neighborhood is shown in Fig. 6.3, and the definition is described as,

$$\text{LBP}(l_c) = \sum_{m=0}^{M} \text{sgn}(l_m - l_c) \times 2^m; \quad \text{sgn}(x) = \begin{cases} 1 & \forall x \geq 0 \\ 0 & \forall x < 0 \end{cases} \quad (6.43)$$

where l_c is the central pixels, and sgn(x) is employed to reserve the sign component between two pixels. The micro-pattern, which is represented as LBP (l_c), is encoded using the sign components in neighborhoods, which is proved to contain most structural information in local regions. In addition, the input signal in HVS is initially extracted from local regions and texture information is perceived in V1 and V2. Therefore, the structure information is believed to be correlated with perceived image quality.

The number of the micro-patterns is $2^8 = 256$, and the histogram is employed to represent the distribution of these micro-patterns, which is described as,

$$H(L) = \sum_{n=0}^{N} f(L, \text{LBP}(l_b)); \quad f(x, y) = \begin{cases} 1, & x = y \\ 0, & \text{otherwise} \end{cases} \quad (6.44)$$

where H is the distribution histogram with the x-axis representing the 256 micro-patterns, $L \in [0256]$, and y-axis denoting the total number of every pattern in image. $f(x,y)$ is a matching function used to compare the code in l_n. And l_n is

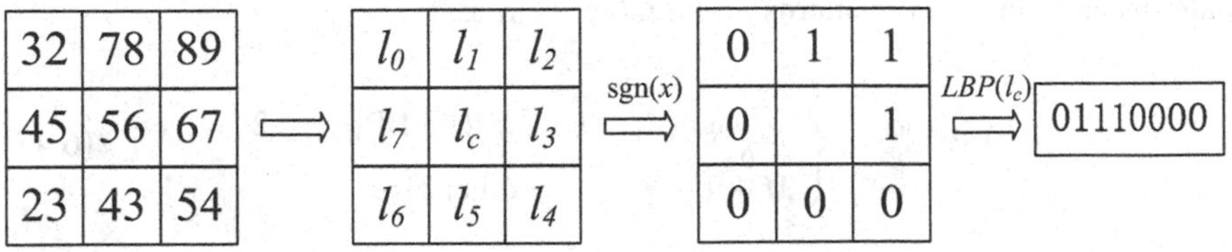

32	78	89
45	56	67
23	43	54

$\Longrightarrow$

l_0	l_1	l_2
l_7	l_c	l_3
l_6	l_5	l_4

$\overset{\text{sgn}(x)}{\Longrightarrow}$

0	1	1
0		1
0	0	0

$\overset{LBP(l_c)}{\Longrightarrow}$ | 01110000 |

Fig. 6.3 Micro-pattern coding

corresponding to l_c in (6.43) with the basic pattern L. When the two patterns are equal, the number of the L-component should plus 1, otherwise remains the same. N is the number of the local regions in image. Since the textures of distorted image are different from those of the reference, the difference between the two texture maps are employed as the quantitative measurement.

There is a continued perfection of LBP. In (Ojala et al. 2002), LBP is completed with the optional parameters of radius and number of the selected neighborhood. And it becomes rotation invariance and uniform to adapt to the texture classification, which is also of great importance in IQA. In order to extract the local structural feature flexibly, the 3×3 neighborhood is replaced with the circularly symmetric neighborhood with parameter r and M. r is the radius of the neighbor, and M is the total number of the pixels selected. Supposing coordinate of the central pixels l_c is $(0, 0)$, the mth neighbor pixel l_m is $(-r\sin(2\pi m/M), r\cos(2\pi m/M))$, and when there is not exact location matched in images, the gray values are estimated by interpolation.

The rotation invariance pattern is defined as,

$$LBP_{M,r}^{ri} = \min\{ROR(LBP_{M,r}(l_c), m)\} \quad m = 0, 1, \ldots, M - 1 \qquad (6.45)$$

where $ROR(x,y)$ is used to move the circular M-bit number x by bit-wise right m times. This method gathers the micro-patterns that have the same rotation pattern to one micro-pattern. And the minimum value of these patterns is assigned to all micro-patterns. In addition, the number of the rotation invariance pattern is 36.

Researchers in (Ojala et al. 2002) find that most extracted patterns are concentrated on certain local binary patterns that reflect the fundamental properties of natural textures. These fundamental patterns are named "uniform" patterns. In "uniform" patterns, $U(x)$ is defined to measure the times of 0/1 translations, shown as,

$$U(LBP_{M,r}) = |sgn(l_0 - l_c) - sgn(l_{M-1} - l_c)| + \sum_{m=1}^{M-1} |sgn(l_m - l_c) - sgn(l_{m-1} - l_c)|$$

$$(6.46)$$

where the coding is regarded as a circle to calculate. And in the experiments in (Ojala et al. 2002), the fundamental patterns are the values of $U(x)$ less than 2. The number of these uniform patterns is 58, and the rest patterns become 1 type. The translation is represented as LBP^{u2}. Therefore, there are 59 patterns in LBP^{u2}. In practice, researchers pay attention to the patterns combined "rotation invariance" patterns and "uniform" patterns and denote them as,

$$LBP_{M,r}^{riu2} = \begin{cases} \sum_{m=0}^{M-1} sgn(l_m - l_c) & if \ U(LBP_{M,r}) \leq 2 \\ M + 1 & otherwise \end{cases} \qquad (6.47)$$

where the total number of LBP^{riu2} is $M + 2$. And the distribution divergence is shown in Fig. 6.4.

Actually, LBP extracting only sign component of textures has a limit application, and the contrast measurement is proposed as,

$$VAR_{M,r} = \frac{1}{M} \sum_{m=0}^{M-1} (l_m - u)^2$$

$$u = \frac{1}{M} \sum_{m=0}^{M-1} l_m$$

(6.48)

where u is the average of gray values of pixel neighborhood, and $VAR_{M,r}$ is the variance of these value, which is related to the contrast in image. Texture analysis is different from image quality assessment. One of the properties of texture analysis is grayscale invariance. However, luminance changes can have an influence on object assessment. Therefore, $VAR_{M,r}$ and LBP are complementary in IQA. And joint distributions $VAR_{M,r}$/LBP can be regarded as a powerful image quality assessment metric.

LBP has been successfully applied on image quality assessment. One of the typical metric is proposed in (Zhang et al. 2013b). The procedure for feature extraction is shown in Fig. 6.5. Firstly, because the density of information is not balanced in the overall image, the multiscale decomposition is employed to extract more structure information at different scales, which is proved to be consistent with responses in V1. The image is decomposed into four scales in (Zhang et al. 2013b) with Laplacian of Gaussian (LOG) filter. Then in each scale, the parameters of LBP algorithm are set as $r = 1, 2, 3$ and $M = 4$. In this case, there are $4 \times 3 \times 6$ binary patterns in each image where 4 refers to the number of scales and 3 is the number of

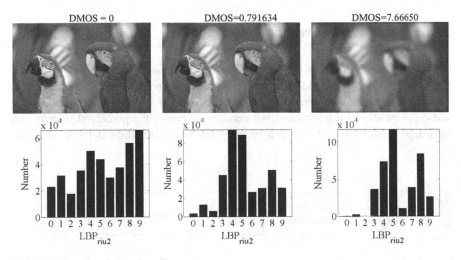

Fig. 6.4 Example of distortion divergence

radii selected. M denotes the number of neighbor pixels selected, and in the meanwhile, there are $M + 2$ rotation-invariant uniform patterns. Therefore, 6 refers to the number of LBP patterns in each scale with r fixed. In the second step, the image is encoded with 72 binary patterns. Finally, the histograms of LBP patterns are normalized and then employed as quality concerned features. The quality mapping procedure is fulfilled with a nonlinear regression model, support vector regression (SVR). With SVR, these quality concerned features are mapped to the subjective scores to train the prediction model. The prediction model can set up the mapping relation between these features and the objective quality scores. Of course, there are many other methods for quality mapping, such as Kullback–Leibler divergence (KLD), chi-square distance, and histogram intersection.

In image quality assessment, some researchers can change LBP to meet special requirements. Zhang et al. employed the histograms with LBP operator as reduced-reference features in (Zhang et al. 2013a). In this experiment, multiscale decomposition is the first step. However, considering that the human visual system is insensitive to the noise under detection threshold, LBP with threshold is proposed. And the thresholds for four decomposition scales are determined as [0.6 4.8 0 0.5]. Since there are many different patterns in LBP operator, the researchers conduct experiments on both LBP$u2$ M,r and LBPriu2 M,r. It is proved that both of them exhibit competitive performance. And the metric with LBP$u2$ M,r is more effective.

Actually, there are some drawbacks about LBP. With gathering the patterns in the overall image, the histogram operator lost the information of spatial location. In addition, the more detailed information is lost by LBP, such as orientation and high-order information, which is important to images. In the meanwhile, the magnitude information is a factor in image. A complete IQA metric taking both the sign and magnitude information into consideration is supposed to reach better performance.

6.3.3 Local Derivative Pattern

With the concatenation of the gradient direction between the central pixel and neighborhood, LBP retained certain texture information, which can be regarded as the first-order circular derivative pattern (Zhang et al. 2010). However, the first-order gradient information is not enough to capture the perceived information for image quality assessment. In a picture, detailed information is more attractive

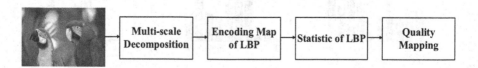

Fig. 6.5 Scheme for LBP algorithm

than background (Costaet al. 2011), and it is necessary to extract the texture features through higher-order derivative information besides the information about the
lines and edges (Ghosh et al. 2007). Therefore, there are many studies on expanding
feature extraction methods in high-order form (Huang et al. 2014; Yuan 2014; Han
et al. 2015). And in order to extract more detailed structural information, high-order
local derivative pattern (LDP) is proposed in (Zhang 2010).

As mentioned above, LBP can extract the nondirectional first-order derivative
features. LDP encodes the certain direction and higher-order features. There are
four directions (0, 45, 90, and 135°) mentioned in LDP, as shown in Fig. 6.6.
Similar to the location in LBP, let l_c be the central pixel shown in Fig. 6.3; then, the
directions are defined as,

$$l'_{0°}(l_c) = l_c - l_4 \tag{6.49}$$

$$l'_{45°}(l_c) = l_c - l_3 \tag{6.50}$$

$$l'_{90°}(l_c) = l_c - l_2 \tag{6.51}$$

$$l'_{135°}(l_c) = l_c - l_1 \tag{6.52}$$

where l_4, l_3, l_2, and l_1 are the right, right-up, up, and up-right gray values of pixels
in l_c's neighborhood, respectively. As shown in Fig. 6.6, for 0°, the difference
between the neighbor of l_c and l_4 is noted in the next square and only record the
sign information for a better representation. The second-order LDP encodes the
values in the next square. Taking direction α as an example, it is defined as,

$$LDP_\alpha^{(2)}(l_c) = \sum_m^M sgn(l'_{\alpha,m} \times l'_{\alpha,c}) \times 2^m \tag{6.53}$$

where function sgn() is the same as in (6.43), and $l'_{\alpha,c}$ and $l'_{\alpha,m}$ are the central
value and mth value in difference square, respectively, which is similar to (6.43).
Thus, the overall direction LDP$^{(2)}$ is generated by concatenating LDP$^{(2)}_\alpha$ in four
directions,

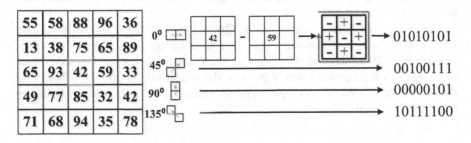

Fig. 6.6 Illustration of second-order LDP

$$\mathrm{LDP}^{(2)}(l_c) = \left\{\mathrm{LDP}_\alpha^{(2)}(l_c)|\alpha = 0°, 45°, 90°, 135°\right\} \qquad (6.54)$$

where $\mathrm{LDP}^{(2)}$, concatenating the codes in all directions in Fig. 6.6, is determined to be a 32-bit code, $\mathrm{LDP}^{(2)} = 01010101001001110000010110111100$. Then, the second-order LDP is extended to dth-order pattern to flexibly extract more detailed features that describe the trend changes of gradient. In a local region, $\mathrm{LDP}^{(d)}$ along four directions is

$$\mathrm{LDP}^{(d)} = \{\mathrm{sgn}(l_\alpha^{d-1}(l_c) \times (l_\alpha^{d-1}(l_1)), \mathrm{sgn}(l_\alpha^{d-1}(l_c) \times (l_\alpha^{d-1}(l_2)),$$
$$\dots, \mathrm{sgn}(l_\alpha^{d-1}(l_c) \times (l_\alpha^{d-1}(l_8))|\alpha = 0°, 45°, 90°, 135°\} \qquad (6.55)$$

where $l\ (d-1)\ \alpha(l_c)$ is the $(d\text{-}1)$th-order derivative in α-direction. The high-order local patterns are believed to provide stronger discriminative detailed texture information than the first-order local pattern in LBP. Some studies have proved that the higher the order is, the more detailed the extracted information is (Zhang et al. 2010; Fran and Hung 2014). Actually, LDP uses the function sgn() to alleviate the noise sensitivity problem. However, there are some pictures about face recognition in (Zhang et al. 2010) showing that over-detailed patterns tend to be sensitive to noise instead of identity information. Therefore, the choice of highest order is made for specific application.

In (Ding et al. 2017), the researchers use multiorder LDP quality-aware features to do image quality assessment and the highest order is determined as 3. The detailed procedure is shown in Fig. 6.7. In order to compare the difference between the reference and distorted images, the multiorder patterns are extracted from the two images.

The purpose of the preprocessing stage is twofold. On the one hand, because images are down-sampled, the extracted features become capable to contain more information. On the other hand, spatial information is retained since the analyzed patches are independent of each other. Then, the multiorder patterns are extracted from each patch. As shown in Fig. 6.7, the first-order patterns can reflect the difference between the reference and distorted images but the second-order and third-order operators can extract orientation and more detailed gradient information. In the meanwhile, with the order increasing, the patterns are contaminated with noises, and Ding et al. determined the highest order as 3. Thus, the multiorder patterns are obtained.

The perceptive feature employed in IQA is the distribution of the multiorder LDP patterns which is measured by the spatial histogram (Zhang et al. 2005; Ahonen et al. 2006) that can retain spatial information than holistic methods. Since the image has been divided into N patches, the spatial histogram $\mathrm{HLDP}^{(d)}(n, \alpha)$ in dth order is calculated in local regions, representing by $R_1,\dots, R_N$, and the definition of $\mathrm{HLDP}^{(d)}(n, \alpha)$ is

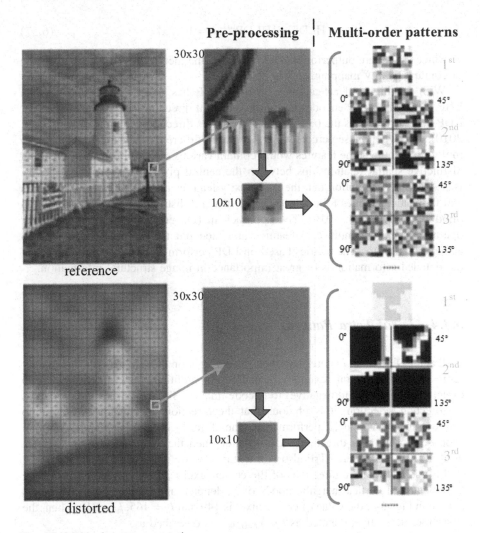

Fig. 6.7 Multiorder pattern extraction

$$\text{HLDP}^{(d)}(n,a) = \{H_{\text{LDP}_\alpha}^{(d)}(R_n)|n = 1,\dots,N;$$
$$\alpha = 0°, 45°, 90°, 135°\}$$

(6.56)

where $H_{\text{LDP}\_\alpha}^{(d)}(R_n)$ is the distribution of the LDP pattern, using (6.44) in α-direction and in local region R_n. And when d = 1, the $H_{\text{LDP}}^{(1)}(R_n)$ is calculated regardless of direction parameters. These $H_{\text{LDP}\_\alpha}^{(d)}(R_n)$ are concatenated to generate the spatial histogram $\text{HLDP}^{(d)}(n, \alpha)$. Then, these features $\text{HLDP}^{(d)}(n, \alpha)$ of different orders are combined into multiorder LDP features, which is described as,

$$\text{HLDP}= \{\text{HLDP}^{(d)}|d = 1,2,3\} \tag{6.57}$$

Since chi-square outperforms other methods, the metric uses chi-square distance to conduct quality mapping.

With the high-order features extracted, LDP has a better performance than LBP. Generally, LBP encodes the binary gradient directions. In contrast, nth-order LDP operator encodes the $(n-1)$th-order derivative direction variations (Zhang et al. 2010). And when these perceptive features from different orders are combined into multiorder features, the features would contain various spatial information encoding distinctive spatial relationships between the central pixel and its neighbor pixels. Based on the assumption that the complexity features are composed of basic units and the natural images follow the general standard distribution, LDP is applied on blind IQA (Du et al. 2016). The approach in (Du et al. 2016) employed LDP operator to extract multiscale features and then put these features into SVR to achieve blind IQA. The metrics based on LDP perform well in IQA and prove that the detailed information is of great importance in image structural information.

6.3.4 Local Tetra Patterns

LBP and LDP extract features based on the distribution of local patterns that mainly represent the structural texture information. In addition, these metrics use two directions (positive and negative) to encode images. Since HVS has the properties of direction selection, it is obvious that the direction information is helpful to improve the IQA models' performance. Therefore, local tetra patterns (LTrPs) are proposed to use four-direction codes to strengthen the extracted features (Murala et al. 2012). Moreover, LTrP also is a texture descriptor for high-order features.

In order to note the directions of the center pixel l_c, there are definitions of the horizontal and vertical neighborhoods of l_c, denoted as l_h and l_v, respectively. As shown in Fig. 6.8, the value of center pixel is 149 and $l_h = 165$, $l_v = 210$. Then, the first-order derivative, denoted as $l^1(l_c)|_{\theta=0,\ 90°}$, is described as

$$l^1_{0°}(l_c) = l_h - l_c \tag{6.58}$$

$$l^1_{90°}(l_c) = l_v - l_c \tag{6.59}$$

and LTrP pattern is proposed to obtain direction information, which is calculated as

$$l^1_{\text{dir}}(l_c) = \begin{cases} 1, & l^1_{0°}(l_c) \geq 0 \ \& \ l^1_{90°}(l_c) \geq 0 \\ 2, & l^1_{0°}(l_c) < 0 \ \& \ l^1_{90°}(l_c) \geq 0 \\ 3, & l^1_{0°}(l_c) < 0 \ \& \ l^1_{90°}(l_c) < 0 \\ 4, & l^1_{0°}(l_c) \geq 0 \ \& \ l^1_{90°}(l_c) < 0 \end{cases} \tag{6.60}$$

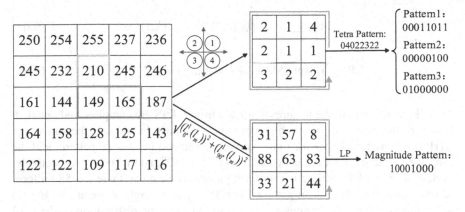

Fig. 6.8 Example of the tetra and magnitude patterns

Therefore, the image is converted into a four-direction mapping. And based on the mapping, the second-order $\text{LTrP}^2(l_c)$ is described as,

$$\text{LTrP}^2(l_c) = \{f_1\left(l^1_{\text{dir}}(l_c), l^1_{\text{dir}}(l_1)\right), f\left(l^1_{\text{dir}}(l_c), l^1_{\text{dir}}(l_2)\right)_1,$$
$$\ldots, f_1\left(l^1_{\text{dir}}(l_c), l^1_{\text{dir}}(l_M)\right)\}|_{M=8} \tag{6.61}$$

$$f_1\left(l^1_{\text{dir}}(l_c), l^1_{\text{dir}}(l_m)\right) = \begin{cases} 0 & l^1_{\text{dir}}(l_c) = l^1_{\text{dir}}(l_m) \\ l^1_{\text{dir}}(l_m) & \text{else} \end{cases} \tag{6.62}$$

And there are four directions of the central pixel. Then, these patterns are separated by the four directions into four parts. For each part, the patterns are converted into three 8-bit binary patterns (Murala et al. 2012). And take the case when the central direction being 1 as example, corresponding to Fig. 6.8, LTrP^2 can be described with the three binary patterns as,

$$\text{LTrP}^2\big|_{\text{Direction}=2,3,4} = \sum_{m=1}^{M} 2^{(m-1)} \times f_2\left(\text{LTrP}^2(l_c)\right)\bigg|_{\text{Direction}=2,3,4} \tag{6.63}$$

$$f_2\left(\text{LTrP}^2(l_c)\right)\big|_{\text{Direction}=\phi} = \begin{cases} 1 & \text{if } \text{LTrP}^2(l_c) = \phi \\ 0 & \text{else} \end{cases} \tag{6.64}$$

where $\phi = 2, 3, 4$.

And the other three parts are also converted to three 8-bit binary patterns. Thus, we get 12(4 × 3) 8-bit binary patterns. This is the structural texture information with the sign component in neighboring. In LTrP (Murala et al. 2012), since the researchers in (Guo et al. 2010) proved that the combination of sign and magnitude components can provide better clues, which cannot be extracted from any individual component, the 13th binary pattern (LP) is proposed to extract the magnitude component with the gradients in horizontal and vertical directions,

$$\mathrm{MA}_{l^1\,(l_m)} = \sqrt{\left(l_{0°}^1\,(l_m)\right)^2 + \left(l_{90°}^1\,(l_m)\right)^2}$$ (6.65)

$$\mathrm{LP} = \sum_{m=1}^{M} \mathrm{sgn}\big(\mathrm{MA}_{l^1\,(l_m)} - \mathrm{MA}_{l^1\,(l_c)}\big) \times 2^{(m-1)}\bigg|_{M=8}$$ (6.66)

As known to us all, the features extracted from LTrPs are complete and detailed. However, these features have a high computational cost. In practice, the patterns in LTrP can be reduced with the uniform pattern, rotation invariance pattern, and the combined pattern which are denoted as LTrP^{u2}, LTrP^{ri}, and LTrP^{riu2}, respectively, as described in LBP. Taking the uniform pattern as example, there are 58 uniform patterns converted from 8-bit patterns. For M neighbor pixels, there are $M(M - 1) + 2$ uniforms. Then, the histogram is used to represent the distribution of these 13 binary-uniform patterns with the histogram defined in LBP (6.44) and the $L\in[0, M (M - 1)+ 2]$.

Since the high-order features are of great importance in image processing, the LTrP algorithm has the higher-order extending,

$$\mathrm{LTrP}^n(l_c) = \big\{f_1\big((l_{\mathrm{dir}}^{n-1}(l_c), l_{\mathrm{dir}}^{n-1}(l_1)\big), f_1\big(l_{\mathrm{dir}}^{n-1}(l_c), l_{\mathrm{dir}}^{n-1}(l_2)\big),$$
$$\ldots, f_1\big(l_{\mathrm{dir}}^{n-1}(l_c), l_{\mathrm{dir}}^{n-1}(l_M)\big)\big\}\big|_{M=8}$$ (6.67)

where l_{dir}^{n-1} is the $(n - 1)$th-order derivatives along horizontal and vertical directions (Murala et al. 2012). Since the order not only provides more detailed information but also more sensitive to noises, the determination of a suitable order is of great importance.

The recent researches on applying LTrP into IQA are presented in (Deng et al. 2017), where the feature extraction is formed by two stages. In the first stage, the visual important regions are enhanced with saliency map, which reflects visual attentions. Distortions in the more attractive regions have larger impact on senses than those in regions with less attraction (Li et al. 2015). Then, the next stage contains two feature extraction methods. With log Gabor filter, the spatial frequency feature is extracted. In the meanwhile, the directional texture analysis is obtained with LTrP operator that makes use of 12(12 = 4 × 3) 8-bit binary patterns.

The statistic of LTrP patterns is removed from the metric. In this case, the patterns are determined as the quality concerned features. There is the similarity measurement on these patterns using chi-square distance to reveal the image quality. Then, the mapping function that synthesizes the results of the similarity measurement into an objective quality score is built with SVR.

LTrP patterns encode the relationship based on directions and magnitude. The extracted information can be applied on blind image quality assessment (Yan et al. 2016). Since LTrP patterns with the precise thresholds are sensitive to noises (Tan and Triggs 2010), and the statistics vary with the noise sensitivity (Zhang et al. 2015). Therefore, the multithreshold is introduced to LTrP to extract more stable and discriminative spatial distribution information. In (Yan et al. 2016), the spatial contrast information is measured with contrast Weber–Laplacian of Gaussian (WLOG) to complement the characteristics of extracted structural information. The final features, which combine the spatial distribution information and spatial contrast information, are synthesized into quality prediction score with the regression function learned by SVR.

The good performance of these metrics proves that the information of space, scale, and orientation are necessary clues in images (Lei et al. 2011). LTrP patterns based on derivative information have the capability of extracting detailed orientation information, which is of great importance in HVS.

6.3.5 Summary

LBP is proposed based on the hypothesis that the complex textures are composed of texture units and the properties of these textures can be represented by the distribution of these units. In fact, this texture information is low-level feature in perceived vision and is essential to image quality assessment. When applying texture information on IQA, we deem that with the image texture altered by distortions, the change of corresponding distribution can reflect the degree of distortion in images. In addition, for blind image quality assessment, there is reference feature derived from the standard distribution of unimpaired images to quantitatively measure the image quality. The good performance of LBP and its variants proves that the extracted texture information is of great importance to IQA. Moreover, the variants extract more discriminative features, which can improve the performance. Therefore, LBP and invariant-based metrics are good descriptors for perceptive image quality. Actually, blind IQA is increasingly important in real-world applications. LBP is a good choice for its relatively high performance and low computational complexity.

However, LBP-based metrics (containing other invariants) also have some drawbacks. Firstly, it is sensitive to noise, and it is proved that the distribution of the patterns can be unstable, which would lead to bad results. Therefore, when applying LBP to IQA, there are some metrics use thresholds to avoid it (Zhang et al. 2013a; Yan et al. 2016). Then, LBP is proposed to code the structural information, regardless of the contrast, which is measured by gradient. Therefore, the complete quantitative measurement of image quality consists of two parts. In (Yan et al. 2016), the authors use WLOG (Weber–Laplacian of Gaussian), which responds to intensity contrast in local regions, to extract local contrast features. In the meanwhile, WLOG is supposed to be more consistent with visual perception. Therefore, this algorithm can be regarded as partly taking the properties of HVS into consideration. Finally, the global distribution of these patterns may loss spatial information in images, and it is proved in (Ahonen et al. 2006; Zhang et al. 2010; Ding et al. 2017) that using spatial histogram or other metrics to retain the spatial information can achieve better performance.

6.4 IQA Based on Independent Component Analysis

Among all the image features, researchers are particularly interest to those of natural scene statistics (NSS) because NSS has shown its superb performance in image processing domain (Simoncelli and Olshausen 2001; Srivastava et al. 2003). Independent component analysis is a very important branch of NSS. As we all known, nature scene signals exhibit striking redundancy and have high dimensionality. Because of this, it is very hard to model the nature scene using a simple model. Independent component analysis is a statistical and computational technique for revealing hidden factors that underlie sets of random variables, measurements, or signals (Hyvärinen et al. 2001). It seems very promising to model the nature signals, as it tries to find hidden independent components in nature signals, and reduces the signal dimensionality to a reasonable size that we can handle. The ICA method is more closed to the low-level visual processing system of animal (Zhang et al. 2012). The localized, oriented band-pass receptive fields learned by independent component analysis are closely resembled those neurons found in the mammalian primary visual cortex (Olshausen and Field 1996).

Independent component analysis has been successfully incorporated into a lot of image processing algorithms such as image denoising, image segmentation, pattern recognition, and image saliency computation (Zhang et al. 2008). In (Kasturiwate and Deshmukh 2009), independent component analysis is used to analyze the ECG signals because huge data is required to evaluate the quality testing in biomedical signal processing field. In addition, it has also been adopted by many recent IQA methods. For example, in Zhang et al. (2012) and Chang et al. (2015), independent component analysis is employed in both two papers as a mathematical tool for extracting the image features. Furthermore, independent component analysis is also used to extract the image features in some stereoscopic image quality assessment methods. For example, an IQA method based on independent component analysis and local luminance consistency of the binocular images is proposed to assess the color stereoscopic in (Geng et al. 2016).

6.4.1 Independent Component Analysis

The generative model in independent component analysis (ICA) is defined as a linear superposition of some random variables (Hyvärinen et al. 2001; Nielsen et al. 2009). Let $\mathbf{I}$ denote the grayscale value in an image patch, according to the ICA generative model, $\mathbf{I}$ can be represented as a linear combination of some bases, which is shown below:

$$\mathbf{I} = \sum_{i=1}^{n} A_i S_i \qquad (6.68)$$

where A_i is called mixing matrix or feature, S_i is one of the base functions learned from nature scenes under certain constraints (e.g., the distributions of S_i should be non-Gaussian). These bases remain constant over image patches. S_i are random variables called latent variables or independent components and are assumed to be statistically independent.

As the generative model is linear, S_i can be derived through an inverse transformation:

$$S_i = \sum_n^{i=1} W_i \mathbf{I} \tag{6.69}$$

where W_i is called separating matrix or feature detector, which can be obtained by inverting the matrix A_i. So, the estimation of A_i is equivalent to determine the values of W_i. In fact, in one image, not all independent components must be extracted. The principal components, including the most important and interesting information, are adequate for analysis. The principal components of an image are defined as

$$\mathbf{F} = a_1 X_1 + a_2 X_2 + \cdots + a_t X_t \triangleq a^T \mathbf{X} \tag{6.70}$$

where $\mathbf{X} = (X_1, X_2, \ldots, X_{t-1}, X_t)$ indicates the pixels in an image, $\mathbf{a} = (a_1, a_2, \ldots, a_{t-1}, a_t)^T$ is the component detector, $\mathbf{a}^T \mathbf{a} = 1$. The variance of $\mathbf{F}$ can be deduced as

$$
\begin{aligned}
\mathrm{Var}(F) &= E(a^T \mathbf{X} - E(a^T \mathbf{X}))(a^T \mathbf{X} - E(a^T \mathbf{X}))^T \\
&= a^T E(\mathbf{X} - E(\mathbf{X}))(\mathbf{X} - E(\mathbf{X}))^T a \\
&= a^T \mathbf{N} a
\end{aligned}
\tag{6.71}
$$

where $\mathbf{N}$ is the matrix of variance in $\mathbf{X}$. Assuming that $\mathbf{U}$ is the feature vector of $\mathbf{N}$ and λ is the characteristic root of $\mathbf{N}$, hence, $\mathbf{N}$ can be defined as

$$\mathbf{N} = \mathbf{U}\lambda \mathbf{U}^T = \sum_{i=1}^{t} \lambda_i u_i u_i^T \tag{6.72}$$

where $\mathbf{U} = (u_1, \ldots, u_t), \lambda = \begin{bmatrix} \lambda_1 & & & \\ & \lambda_2 & & \\ & & \ddots & \\ & & & \lambda_t \end{bmatrix}$. Hence, the variance of $\mathbf{F}$ can be continually deduced as

$$
\begin{aligned}
\mathrm{Var}(\mathbf{F}) &= a^T \mathbf{N} a = \sum_{i=1}^{t} \lambda_i (a^T u_i)(a^T u_i)^T = \sum_{i=1}^{t} \lambda_i (a^T u_i)^2 \le \lambda_1 \sum_{i=1}^{t} (a^T u_i)^2 \\
&= \lambda_1 (a^T \mathbf{U})(a^T U)^T = \lambda_1 a^T \mathbf{U}\mathbf{U}^T a = \lambda_1 a^T a = \lambda_1
\end{aligned}
\tag{6.73}
$$

The variance of $\mathbf{F}$ obviously cannot exceed λ_1, so the maximum variance is λ_1, i.e., Var $(uT\ i\mathbf{X}) = \lambda_i$. Moreover, the correlation of principal components can be defined as

$$\text{Cov}(u_i^T X, u_j^T X) = u_i^T N u_j^T = \sum_{\alpha=1}^{t} \lambda_\alpha (u_i^T u_\alpha)(u_\alpha^T u_i) = 0 \qquad (6.74)$$

It shows that components are not correlated with each other. This property is needed for independent component extraction. As $\lambda_1 \geq \lambda_2 \geq \ldots \geq \lambda_t > 0$, the components also named feature vectors are sorted according to the sequence of the characteristic roots λ. Kicking out the components with small characteristic roots, the residual components are the principal components as they have higher ratio in the components.

Independent component analyzer adopts an orthogonal transformation in feature extraction. The aim of ICA is to make the transformed output as much as possible statistically independent of the various signal components, thus to estimate the basic structure of space or the source signal in the mixed signals observed (Guo and Zong 2012). The way to guarantee independence is gradually expanding sparseness of the outputs from image feature detecting. A feature detector is defined as

$$S_i = W_i \mathbf{F} \qquad (6.75)$$

where W_i is independent feature detector, S_i is an independent feature, and F is the principal component. Because the latent variables are assumed to be statistically independent, the joint probability mass function is therefore the multiplication of the probability mass function of each random variable (assuming there are n latent variables, denote each distribution function as p_i),

$$p(S_1, \ldots, S_n) = \prod_{i=1}^{n} p_i(S_i) = |\det(W)| \prod_{i=1}^{n} P_i(W_i^T \mathbf{F}) = |\det(W)| \prod_{i=1}^{n} P_i\left(\sum_{j=1}^{n} W_{ij}F_j\right)$$

$$(6.76)$$

where $p\ (S_1, \ldots, S_n)$ is PDF of independent features, and n principal features are chosen to detect the independent features. Maximum likelihood estimation (MLE) (Nielsen et al. 2009) is applied to estimate the model parameters by maximizing the log-likelihood of the joint distribution probability. Log-likelihood, which serves as a measure of sparseness, is defined as

$$\log L(W_1, W_2, \ldots W_h) = h \log|\det(\mathbf{W})| + \sum_{i=1}^{n} \sum_{i=1}^{h} \log P_i(W_i^T \mathbf{F}) \qquad (6.77)$$

where h is feature vector extracted by independent components analysis. $\mathbf{W}$ can be estimated by maximizing this log-likelihood:

$$\hat{\mathbf{W}} = \max_{W} \arg \log L(W_1, \ldots, W_h) \tag{6.78}$$

Since the components are sorted according to the sequence of the characteristic roots λ, the independent components are uniform regardless of the order of image patches, as illustrated in Fig. 6.9. Compared with the raw image, its patches in random order seem to have the same independent components. It indicates that independent analyzers are less sensitive to some disturbance that changes the structure of the raw image.

Once we get the estimated filter set $\mathbf{W}$, we can decompose the image by applying each filter on it and generate a set of independent components. Each component can be modeled separately because they are deemed to be statistically independent (Zhang et al. 2012). The marginal distribution of independent components can be well-fitted with the GGD model. Figure 6.10 shows the distribution of a latent variable and its GGD fitting. And the fitting was done using maximum likelihood estimation.

General Gaussian distribution (GGD) (Wang and Simoncelli 2005) model is used to parameterize the distribution of all independent components and generate n pairs of parameters, which are denoted as (α_1, β_1), ..., (α_n, β_n). These are features we extracted from images. Features that extracted from the perfect images are served as reference. Once the reference features are calculated, they remain constant. Features of the distorted images are also extracted. Image quality can then be calculated by measuring the variation of these features. The Kullback–Leibler distance (KLD) is employed to measure the difference between the reference images and the distorted images. Let $q_i(x)$ represent a filter response of nature scene, and the distribution of the independent component of distorted image is denoted as $p_i(x)$. Then, KLD between the two features can be written as,

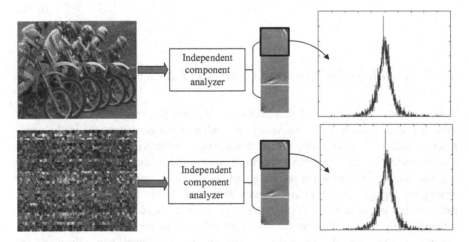

Fig. 6.9 Raw image and its patches in random order under the independent component analyzer

Fig. 6.10 Marginal
distribution of a latent
variable with its
corresponding GGD fitting.
The solid line represents the
histogram of the latent
variable; the dashed line
represents the fitting

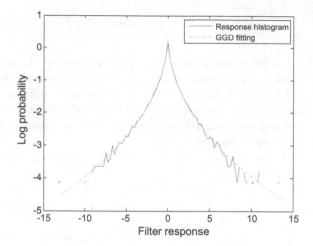

$$d_i = D(p_i(x)\|q_i(x)) = \int p_i(x) \log\frac{p_i(x)}{q_i(x)}dx \qquad (6.79)$$

where

$$q_i(x) = \frac{\beta_i}{2\alpha_i\Gamma(1/\beta_i)} e^{-(|x|/\alpha_i)^{\beta_i}}.$$

For an image, after calculating KLD of all features, the distance d_i is utilized to obtain a distortion index.

$$D = \frac{\sum\limits_{i=1}^{n} d_i}{n} \qquad (6.80)$$

6.4.2   Topology-Independent Component Analysis

To introduce a topographic representation in the independent component analysis model, it is necessary to define which are independent or dependent in the topographic structure. Topography means that the cells in the visual cortex are not in a random order. Instead, they have specific spatial organization. When moving on the cortical surface, the response properties of the neurons change in systematic ways among the topographic structure (Meyer-Bäse et al. 2003). In the statistical model, neurons that are near to each other are considered to have statistically dependent outputs, while neurons that are far from each other are believed to have independent outputs. According to this property of visual system, the linear feature detectors are arranged on a grid so that any two feature detectors that are close to each other have dependent outputs, while feature detectors that are far from each other have independent outputs.

Topographic order of the independent components is correlated with the location of the respective parts. The definition of topographic order can be accomplished by a function $h (i, j)$, which expresses the proximity between the ith and jth components. Neighborhoods can thus be defined as one-dimensional or two-dimensional; 2D neighborhoods can be square or hexagonal (Hyvärinen et al. 2001). Usually, the neighborhood function is defined as a monotonically decreasing function of some distance measure, which implies that it is symmetric: $h (i, j) = h (j, i)$, and has constant diagonal: $h (i; i) = const$. For all I, the neighborhoods consideration can be defined by

$$h(i,j) = \begin{cases} 1, & if\ |i-j| \leq m \\ 0, & otherwise \end{cases} \quad (6.81)$$

where $h (i, j)$ indicates whether the jth feature extracted is around the ith features or not, and m is the neighborhood size. Secondly, we define the "local energy" is

$$E_i = \sum_{j=1}^{m} h(i,j)S_j^2 \quad (6.82)$$

Besides, they are dependent on each other in the neighborhood. These components can be combined together to create a new ingredient; thus, the new components are independent of each other. The combination can be defined as

$$e_i = \sqrt{S_i^2 + \sum_{j=1}^{k} h(i,j)S_j^2} \quad (6.83)$$

where e_i indicates the ith components extracted by the new independent component analyzer, and k components are around ith feature. In this way, the new independent component analyzer eliminates the topographic order to extract more independent components.

In general, there are two kinds of neighborhood sizes to determine the topographic order. The first one is every neighborhood consisted of a 3×3 square of nine units. The second one was defined similarly, but this time with a 5×5 neighborhood (Wei and Li 2010). With this bigger neighborhood, the topographic order is more strongly influenced by orientation. 3×3-neighborhood topographic order is illustrated in Fig. 6.11. Just as the definition in (6.82), the eight components around S_i are dependent on each other, so all of the nine components can be combined together in the form of "local energies" defined in (6.83) to create a new component. All the created new components are independent of each other.

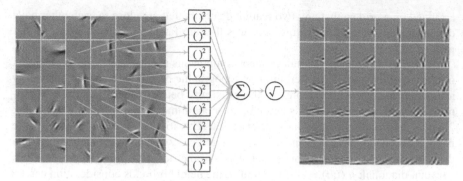

Fig. 6.11 Topographic order of 3 × 3 neighborhood

There is a training process before assessing the quality of distorted image. A training set, which consists of many raw naturalistic images, should be analyzed by the new topographic independent components analyzer. The images are divided into 16 × 16 patches. These patches then experience three steps to extract independent components. They first are preprocessed to extract the principal components to simplify the independent component extraction. The principal components are sorted for the next step—estimated independent components extraction. Thirdly, topographic order consideration eliminates the high-order dependence in the estimated independent components. The new independent components are fitted to a Gaussian model. Such distribution treated as an ideal distribution has a measurable deviation compared with that of independent components extracted from the distorted image. These measurable deviations are used as the measure of the quality of the distorted image. This whole framework is illustrated in Fig. 6.12. The outputs of the training can be applied to any images in database, so it is not necessary to train the model every time before predicting the quality.

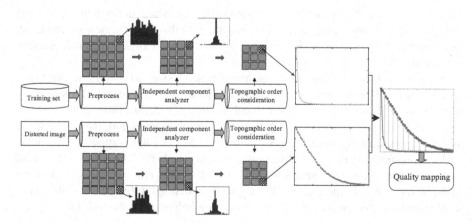

Fig. 6.12 Framework of IQA based on TICA

In order to predict the quality of distorted images, a relationship should be established between the deviations and the quality prediction. The Gaussian density is employed to approximate the histogram of e_i to form an ideal distribution as the fixed feature e_i can well present the inner property in the pristine images. The Gaussian density is defined as

$$p(e_i) = \frac{1}{c}\exp(-\frac{|e_i|^\alpha}{b^\alpha})$$ (6.84)

where $b = \sqrt{\Gamma(\frac{1}{\alpha})/\Gamma(\frac{3}{\alpha})}$, $c = \left[2b\sqrt{\pi}\Gamma(\frac{1}{\alpha})\right]/\left[\alpha\Gamma(\frac{1}{2})\right]$, and $\Gamma(\cdot)$ is the gamma function, i.e., $\Gamma(c) = \int_0^\infty t^{c-1}e^{-t}dt$.

KLD is widely applied in the prediction of IQA to quantify the difference accurately (Liu et al. 2010). Assuming that there are M fixed features, i.e., $e = (e_1, ..., e_M)$, M numbers of KLD are involved in one image. Since different topographic structures have different distortion attributions, a reasonable combination of KLD is needed. This combination of KLD in one image is directly defined as

$$D = \sum_{i=1}^{M} f_i \times \int p_i(x) \log\frac{p_i(x)}{q_i(x)}dx$$ (6.85)

where D is the total KLD of one certain image, and f_i indicates the ratio of the ith KLD in D. $q(x)$ is probability density function of the ideal distribution, and $p(x)$ is the probability density function of the distorted image histogram.

6.4.3 Independent Subspace Analysis

It is probably quite safe to say that independent subspace analysis (ISA) is for complex cells while ICA is for simple cells (Ding et al. 2016a). When estimated from natural image data, in ISA, dependencies of linear features are considered, which is similar to what is computed by complex cells in primary visual cortex (Mittal et al. 2012). Therefore, ISA takes the advantage of the complex cells in V1 to extract features close to the inherent image quality which is definitely consistent with the visual perception.

Like in the case of ICA, firstly images are preprocessed with three steps: (i) natural images are divided into numerous patches of size 32×32 to form vectors of 1024 dimensions; (ii) all the patches undergo a contrast gain controlled by regional normalization; (iii) normalized patches are processed by principal component analysis, reducing the dimensions to 256. Then, linear features (l_i) are extracted from whitened data that are obtained by normalizing the variances of these 256 principal components,

$$l_i = \sum_{x,y} v_i(x,y) W(x,y) \tag{6.86}$$

where v_i is the linear feature detector which is orthonormalized, and W is the whitened data calculated by combining linear transformation and principal component analysis. It is, of course, important to define a function to compute the strength of a subspace feature which consists of a number of linear features. The square root of the sum of the squares is proved to be a useful measure to approaching this. Thus, the nonlinear features take the form as,

$$N_k = \sqrt{\sum_{i \in L(k)} l_i^2} \tag{6.87}$$

where $L(k)$ is the kth subspace and N_k is the kth nonlinear feature.

So, it makes sense to determine how many dependent linear features in one subspace by minimizing the distance of an image from the best approximation that the subspace feature is able to provide

$$\min_{l_i, i \in L(k)} \left(\sum_{x,y} I(x,y)^2 - \sum_{i \in L(k)} l_i^2 \right) \tag{6.88}$$

where $I(x, y)$ is an image. The number of feature subspace is determined according to the predefined subspace size. In this letter, the subspace size is 4, which means there are 64 = 256/4 subspaces for a natural image and each subspace forms a nonlinear feature N_k.

Furthermore, expanding sparseness of features can increase the independence of the nonlinear features. Maximization of the sparseness is equivalent to estimation of the statistical generative model by maximization of likelihood (Mittal et al. 2012). Here, the log-likelihood is given by

$$\log L(v_1, \cdots, v_n) = T \log|\det(V)| + \sum_k \sum_{t=1}^{T} h(N_k^2) \tag{6.89}$$

where V is the set of feature detectors $(v_1..., v_n)$, T is the number of independent subspaces, and h is the nonlinear function adopted in the likelihood or sparseness measure as a smoothed version of the square root. Thus, the linear features in one subspace are dependent and the nonlinear features representing each subspace are independent. As shown in Fig. 6.13, features a and b, which are chosen in a same subspace, almost show the same representation of grayscale map. Evidently, they follow the similar distributions in their histogram. By contrast, in a different independent subspace, feature c has a different representation of grayscale map and

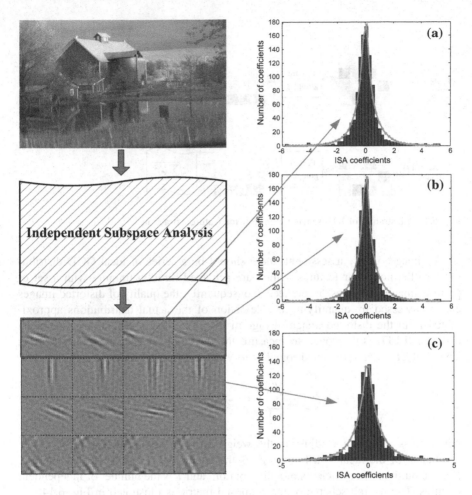

Fig. 6.13 Distribution of features

follows a dumpy distribution. It proves that the nonlinear features are independent, but the components (i.e., linear features) in the same subspace are not.

It has been demonstrated that sub-band and responses of natural scenes tend to follow a non-Gaussian (heavy-tailed) distribution; thus, we use a generalized Gaussian distribution (GGD) model to characterize the nonlinear features,

$$p(N_k) = \frac{1}{c}\exp\left(-\frac{|N_k|^{\alpha}}{b^{\alpha}}\right) \tag{6.90}$$

where $b = \sqrt{\Gamma(\frac{1}{\alpha})/\Gamma(\frac{3}{\alpha})}$, $c = \left[2b\sqrt{\pi}\Gamma(\frac{1}{\alpha})\right]/\left[\alpha\Gamma(\frac{1}{2})\right]$, and $\Gamma(\cdot)$ is the gamma function, i.e., $\Gamma(s) = \int t^{s-1}e^{-t}dt$.

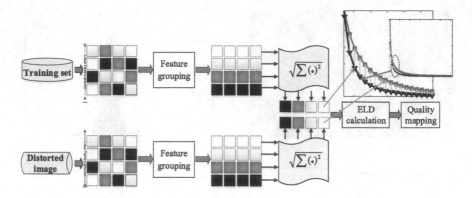

Fig. 6.14 Illustration of IQA scheme with ISA for feature extraction

For image quality assessment, there should be a series of ideal GGD distributions of the nonlinear features, which are achieved through training with a set of pristine images, serve as "reference." Consequently, the quality of distorted images can be assessed by quantifying the deviation of the actual distributions approximated from the distorted image histogram from the ideal distributions. Euclidean distance (ELD) is employed to measure the deviation of each nonlinear feature. Then, all ELDs are combined together to yield a distortion index,

$$Q = \sum_{k=1}^{T} \lambda_k \sum_{j=1}^{R} \left(\beta_2^{(j)} - \beta_1^{(j)} \right)^2 / \left(\beta_2^{(j)} \beta_1^{(j)} \right) \tag{6.91}$$

where λ_k is a factor for adjusting the weight of each nonlinear feature, β_1 is the coefficient of the ideal distribution, β_2 is the coefficient of the actual distribution, R is the number of coefficients of a distribution, and T is the number of independent features. The overall scheme of the proposed metric is illustrated in Fig. 6.14.

6.4.4 Kernel-Based Nonlinear Independent Component Analysis

ICA decomposes images by a linear transformation as mentioned above. However, generally, it is a significant challenge for such method to deal with nonlinear problems due to the unique and complicated mathematical representation in coded digital images (Ding et al. 2016b). Kernel independent component analysis (Kernel ICA or KICA) is proposed to overcome the disadvantage of ICA (Yu et al. 2014). Minimizing mutual information in all nonlinear function space is employed in kernel method (Ding et al. 2016a). Kernel method or kernel trick is widely employed in learning and optimizing algorithms as a nonlinear similarity measure

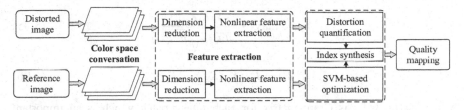

Fig. 6.15 Illustration of IQA scheme with ISA for feature extraction

which makes it possible to apply linear methods to vectorial and nonvectorial data and enable the nonlinear feature extraction. This whole scheme is illustrated in Fig. 6.15.

High dimensions can provide lager number of variables for observing, but it also presents many challenges for us. Getting rid of some redundant information can help us calculate well. To keep significant information as much as possible, the original images are usually projected into low-dimensional space. Principle component analysis is usually used to reduce the dimensions. To a certain extent, KICA is one of the efficient dimension reduction methods (Miyabe et al. 2009). Like in the case of ICA, principle component analysis is employed as the preprocessed tool to reduce the dimension. The aim of ICA is to find statistically independent features which can form the variables by linear combination. With respect to the KICA method, before linear operation, nonlinear transformation is conducted by mapping the input data into an implicit reproducing kernel Hilbert space (RKHS),

$$\phi : x \in \mathbb{R}^n \rightarrow \phi(x) \in F \tag{6.92}$$

where x represents the input image patches, F denotes an RKHS on $\mathbb{R}$, $\phi(x) = K(\cdot, x)$ stands for the mapping way, and $K(\cdot, x)$ represents a kernel function. To get the statistical independent components, F-correlation between the two features obtained from the same content image is calculated as follows:

$$\rho_F = \max_{f_1, f_2 \in F} \mathrm{corr}(f_1(x_1), f_2(x_2)) = \max_{f_1, f_2 \in F} \frac{\mathrm{cov}(f_1(x_1), f_2(x_2))}{\sqrt{\{\mathrm{var}(f_1(x_1))\} \cdot \{\mathrm{var}(f_2(x_2))\}}} \tag{6.93}$$

where $f_1(x_1)$ and $f_2(x_2)$ represent two features from RKHS of the input data. And the constant function is defined as:

$$I_{rF} = -\frac{1}{2}\log(1 - r_F) \tag{6.94}$$

The value of IrF is always nonnegative. In addition, if and only if it equals to zero, the variables are independent of each other.

Based on the above description, estimating and minimizing F-correlation is used to separate individual nonlinear components in KICA method,

$$\hat{\rho}_F = \max_{\alpha, \beta \in \mathbb{R}^m} \frac{\alpha^T K_i K_j \beta}{\sqrt{\alpha^T K_i^2 \alpha} \sqrt{\beta^T K_j^2 \beta}} \qquad (6.95)$$

$$C(W) = \hat{I}_{\rho F}(K_1, K_2, \ldots, K_m) \qquad (6.96)$$

where $K_i (i \in \mathbb{N}^+)$ is Gram matrix for each image patch which is an important application to compute linear independence in nonlinear space. Then, the demixing matrix W from observed image data is calculated. Since the total number of components derived by ICA is determined by number of image patches, the components are too much to be selected because of their equivalence. The difficulty and calculation cost reduced greatly by calculating the inner products between images in feature space rather than in original space.

Generally, the distortion index can be calculated by measuring the difference between the features extracted from reference and distorted images. It is not obvious for deviations of the distributions of different kernel components due to the robustness of KICA. Therefore, traditional methods such as KLD and ELD are hardly measure the deviations. For an example, the distributions of kernel-independent components of a sample are demonstrated in Fig. 6.16, where the distribution with blue is derived from a reference image (named "monarch.bmp" in LIVE database), the red derived from its distorted version contaminated by serious fast-fading, the rest one is from another content image (named "ocean.bmp" in LIVE database). It is obvious that these three feature distributions have almost the same probability distribution although they are with different degree of distortions even with different contents. Some researchers proposed to use correlation coefficient to quantify the deviations (Ding et al. 2016b),

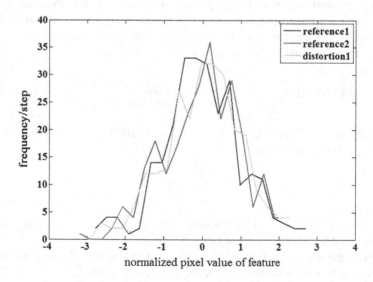

Fig. 6.16 Distribution of a sample kernel-independent component

$$\text{corr}(C_i(x,y)) = \frac{\sum\limits_{j=0}^{m}(x_j - \bar{x})(y_j - \bar{y})}{\sqrt{\sum\limits_{j=0}^{m}(x_j - \bar{x})^2 \cdot \sum\limits_{j=0}^{m}(y_j - \bar{y})^2}} \tag{6.97}$$

where x and y are the extracted features from the reference image and distorted image, respectively, and m represents the number of image patches.

6.4.5 Summary

Nowadays, researchers are attempting to reveal the statistical characteristics within the natural scenes, which are less subjective and drive more applications of statistical tools and methods (Zhang and Chandler 2013; Ding et al. 2014). It is hypothesized that distortion would disrupt the normal statistical property of original images regardless of distortion types. Based on this principle, statistical analysis is superior to the previous methods. Furthermore, obviously, the more consistent IQA algorithm is with the human visual cortical feature detectors, the better the model performs. From the viewpoint of information theory, it is claimed that human visual cortical feature detectors might be the final result of a "redundancy reduction" process, in which the activation of each feature detector is supposed to be statistically independent of each other (Simoncelli and Olshausen 2001). If the extracted features are independent enough, the predicted quality of distorted images is similar to that perceived by a human (Simoncelli 1997). So, a more marvelous independent analyzer is needed for better performance in IQA methods.

In the recent years, researchers pay more attention to ICA which decomposes the input image signal into several independent components to find the most independent components by a linear transformation. ICA seems to be a powerful tool for image feature extraction because it discloses the similarities between statistical properties of natural scenes and the characteristics of HVS (Olshausen and Field 1996). Moreover, ICA has been successfully applied in many fields, such as brain disorder research and pattern recognition (Liu and Yang 2009). However, in most cases, the components considered to be independent are clearly dependent on each other. Usually, it is useful to estimate this dependence structure. Such high-order dependence keeps those models from extracting features independent enough. Topology independent component analysis (TICA) abandons the shortcomings of ICA that disposes the data regardless of the topographic relationship among the output data.

Because it has been claimed that the features extracted by ICA are not completely independent on each feature but independent on each group, another new method ISA is also proposed to address the shortcomings of ICA. Different from the simple cell simulated by conventional ICA, independent subspace analysis (ISA) is employed to group the dependent features into a complex cell using a nonlinear transformation (Jiao et al. 2013). The resulting model gives nonlinear features which turn out to be very similar to complex cells in visual system when the parameters are learned from natural images. Clearly, the nonlinear features are more independent than linear features. Evaluations demonstrate that it is a much insightful way for image inherent quality measurement, and most importantly, it can be applied to different distortion types. In addition, kernel method which transforms the linear decomposition problem into a nonlinear one is also proposed to overcome the disadvantage of ICA.

6.5 IQA Based on Multifractal Analysis for Feature Extraction

Most of the natural objects are so irregular and complex that they cannot be described by ideal shape primitives, such as cubes, cones. According to the concept of self-similarity, (Mandelbrot 1967) proposed the fractal geometry of nature. Fractal geometry is based on the assumption that all space scales are jointly and equally important and that the key information lies in the mechanisms relating them to each other, which explains why fractal is also scaling invariance. Fractal geometry provides a mathematical model for many complex objects found in nature, such as coastlines, mountains, clouds (Mandelbrot and Wheeler 1983; Pentland 1984; Peitgen et al. 2004). Many researchers have applied the fractal theory to several fields of scientific research, such as biology (Mirny 2011; Provata and Katsaloulis 2010), physics (Loh et al. 2012; Zhou 2008), and medicine (Humeau et al. 2010; Stosic and Stosic 2006), where the characteristic of self-similarity and scale invariance are found critically important (Field 1987). (Pentland 1984) noticed that the fractal model of image could be used to acquire shape information and to distinguish smooth or rough textured regions, which is in accordance with human perception. Therefore, multifractal analysis is a powerful tool to extract significant image features for image understanding.

Digital images are generally highly structured, and the complexity of texture is crucially important to human perception. Texture irregularity and surface complexity are fundamental features of nature scenes, such as the edge and contour information of images. It is difficult to quantify the surface morphology because of the difficulty of describing the surface mathematically. However, it becomes possible to capture the complexity information of an image based on the fractal theory. Furthermore, these properties are associated with the visual perception of an image. In other words, fractal geometry established a bridge between mathematical model of a digital image and the subjective visual distinction. Therefore, fractal is becoming a powerful and efficient tool to characterize and describe digital images.

Furthermore, to satisfy both accuracy and efficiency of IQA metrics, a novel method of image quality-related feature extraction based on multifractal is proposed (Zhang et al. 2014). Multifractal analysis is put forward to extract the statistical complexity information of images which is in accordance with HVS. Beyond fractal theory, multifractal analysis characterizes how globally irregular a scene is. Multifractal spectra concentrate on describing the fluctuations along space of the local regularity of an object. In the context of image quality assessment, distortions may change the multifractal spectra of original image as well as perceived image quality. The visual quality degradation can be evaluated by quantifying the discrepancies of multifractal spectra of reference images and those of the distorted images.

6.5.1 Fractal Dimension

Self-similarity is an essential property of fractal in nature and may be quantified by a fractal dimension (FD). As is well known, FD is a concept related to certain key features of fractals: self-similarity and scale invariance, which is powerful in analyzing the irregularity of natural images (Ding et al. 2015). Recently, FD has been widely applied in texture analysis and segmentation (Chaudhuri and Sarker 2002; Liu and Chang 1997; Ida and Sambonsugi 1998), shape measurement and classification (Neil and Curtis 1997), and so on (Lin et al. 2001; Asvestas et al. 1998).

Different methods have been proposed to estimate FD which can be classified into three major categories: the box-counting methods, the variance methods, and the spectral methods (Russell et al. 1980; Balghonaim and Keller 1998). The box-counting dimension is the most frequently used for measurements in various application fields due to its power in representing the complexity of images and its easy implementation (Mandelbrot and Wheeler 1983; Li et al. 2009). Among the practical box-counting methods brought forward for FD estimation, the differential box-counting (DBC) method is considered as a better method (Sarker and Chaudhuri 1994; Yu et al. 2005).

As a start, a digital image of $N \times N$ pixels is mapped onto a three-dimensional spatial surface with (x, y) representing the pixel position on the image plane and z denoting intensity/gray level of pixel. For example, Fig. 6.17 shows the

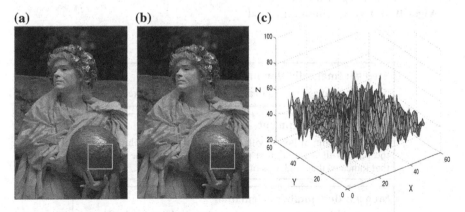

Fig. 6.17 Three-dimensional surface of pixel gray level

three-dimensional surface of gray level of a selected image patch, where (a) is a color image, (b) is the gray-level image converted from (a), and (c) is the three-dimensional surface of a selected image patch.

The basic principle of box dimension is to cover a binary image with a series of boxes (size = $s \times s$). The scale of each box is $r = s$, where $N/2 \geq s > 1$ and s is an integer (Sarker and Chaudhuri 1994; Yu et al. 2005). The procedure is shown as Algorithm 1. If these boxes completely cover the subject and when the size approximates to zero, the ratio of the logarithm of the number of boxes to the reciprocal logarithm of size is referred to as box dimension. By covering a signal with boxes of length r, the fractal dimension is estimated as the negative slope of logarithmic N versus logarithmic r.

$$FD = \lim_{r \to 0} \frac{\ln(N(r))}{(1/r)} \tag{6.98}$$

where $N(r)$ is some kind of complexity measure and r is the scale which the measure is taken. The fractal dimension represents the irregular degree of an image, and a larger FD refers to more complex and rough textures in the image. The (x, y) space is partitioned into grids of size $s \times s$. On each grid, there is a column of boxes of size $s \times s \times s'$. Let the minimum and maximum gray levels of the image in the (i, j)th grid fall in box number k and l, respectively. In DBC approach, $n_r (i, j) = l - k + 1$ is the contribution of $N (r)$ in the (i, j)th grid. As an example, $n_r (i, j) = 3 - 1 + 1$ where $s = s' = 3$. Taking contributions of all grids,

$$N(r) = \sum_{i,j} n_r(i,j) \tag{6.99}$$

A larger fractal dimension refers to more complex and rough textures in an image that is in accordance with visual perception. While fractal dimension is efficient in describing the irregularities of natural scenes, some related observations noticed that fractal dimension may lose some local complexity information during the calculation (Ding et al. 2015).

Algorithm 1 Box-counting method

Step #1: Image division/partition
divide a test image into equivalent patches of $N \times N$ pixels

---

Step #2: Box assignment
assign a column of boxes with size of $r \times r \times s$ on a block starting the pixel with the minimum gray level, where s is the box height that indicates the roughness of an image.

Step #3: Box number calculation
calculate the box number based on the maximum and minimum gray levels of each block.

6.5.2   Multifractal Spectrum

Despite the efficiency of fractal dimension in measuring the complex features of texture, it should be noted that images with different content may have the same fractal dimension (Allain and Cloitre 1991). In order to extract more fractal information of images, multifractal, which can be seen as an extension of fractal, is introduced into image quality-related feature extraction for a more effective fractal description of nature scene (Zhang et al. 2014; Ding et al. 2015). Multifractal spectra describe the evolution of the probability distribution of fractal structures.

In the box-counting-based multifractal analysis, the (x, y) plane is partitioned into nonoverlapping boxes of size $r \times r$. Let define scale $\varepsilon = r/N$. $P_{ij}(\varepsilon)$ is the deposition probability of the box (i, j) in scale ε,

$$P_{ij}(\varepsilon) = \frac{h_{ij}}{\sum h_{ij}} \tag{6.100}$$

where h_{ij} is the sum of all pixel intensities contained in box (i, j) and ε is scale measure computed by the ratio of box size to the whole image size.

After the calculation of P_{ij}, the partition function $\chi_q(\varepsilon)$ is introduced and it is expressed as a power law of ε with an exponent $\tau(q)$,

$$\chi_q(\varepsilon) = \sum P_{ij}(\varepsilon)^q = \varepsilon^{\tau(q)} \tag{6.101}$$

where q is the moment order. It is clear that when $q \gg 1$, the large probability subset will be the main part of $\chi_q(\varepsilon)$ and when $q \ll 1$, the small probability subset will be the main part of $\chi_q(\varepsilon)$. Thus, the delicate inner structure of the fractal set can be analyzed. This power law constitutes the fundamental relation connecting the concept of fractal. The exponent $\tau(q)$ can be obtained from the slope of $\ln\chi_q(\varepsilon)$–$\ln\varepsilon$ curve. The multifractal spectra $D(h)$ are computed by performing a Legendre transformation as follows

$$h = \frac{d[\tau(q)]}{dq}. \tag{6.102}$$

$$D(h) = hq - \tau(q) \tag{6.103}$$

where h is defined to be the singularity of the subset of probabilities and $D(h)$ be the fractal dimension of h subset.

The procedure for multifractal spectra calculation of selective patch in image "Monarch" is illustrated in Figs. 6.18 and 6.19, where the color image shown in Fig. 6.18a is first transformed into grayscale image shown in Fig. 6.18b. Then, we decompose the selective patch into small boxes of different scales and compute the deposition probability of every box in each decomposition level to get partition function as shown in Fig. 6.19a. The exponent $\tau(q)$ is calculated by partition

(a)   **(b)**

Fig. 6.18 Image "Monarch" and Multifractal the selected patch in it. **a** the color image, **b** the corresponding gray-level image

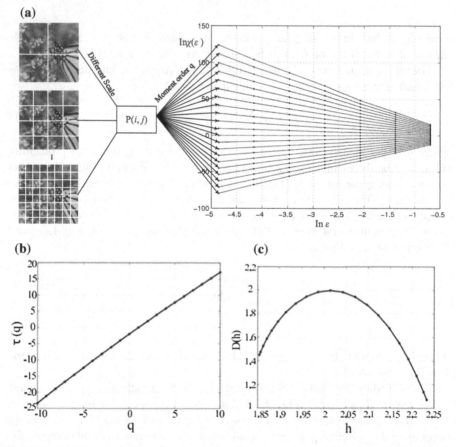

Fig. 6.19 Multifractal analysis of selected patch in image "Monarch". **a–c** computation of logscale diagram $\ln\chi_q(\varepsilon)$ versus $\ln(\varepsilon)$, $\tau(q)$ versus q, and multifractal spectra $D(h)$

function in the following which is represented in Fig. 6.19b. Finally, through Legendre transformation, the multifractal spectra as depicted in Fig. 6.19c are obtained.

As expected, various distortions may inevitably affect the multifractal spectra so that makes them different from the spectra extracted from the original image. For example, blur distortion makes an image smoother while block-based compression causes an image to be more sharp and complex. All these characteristics will eventually change the multifractal spectrum. Experiments demonstrate that the multifractal spectra of images with different distortion are quite different from those extracted from the original image. As shown in Fig. 6.20, (a) is the reference image "bikes" from LIVE database, and (b)–(d) are distorted versions of "bikes" with different DMOS. DMOS is the differential mean opinion score of subjective image quality assessment, where a smaller DMOS refers to a better perceptual image quality. And the multifractal spectra of distorted "bikes" images with different distortions are presented in Fig. 6.21, where the black curve is the multifractal spectra of the selected patch in the reference image shown in Fig. 6.20a, and the blue triangle, red square, and green circle are the multifractal spectra corresponding to the selected patches in distorted images Fig. 6.20b, c, and d, respectively.

Obviously, the multifractal spectra in blue triangle are most similar to those of the reference image in black, which indicates image shown in Fig. 6.20b has good image

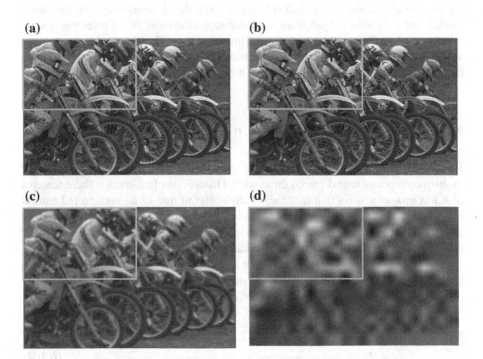

Fig. 6.20 Different distortion levels of the image "bikes" from LIVE. **a** The reference image; **b** "bikes" contaminated by JPEG2000, DMOS = 38.4690; **c** Gaussian blur distortion of "bikes", DMOS = 55.8753; and **d** Fast-fading distortion of "bikes", DMOS = 74.0254

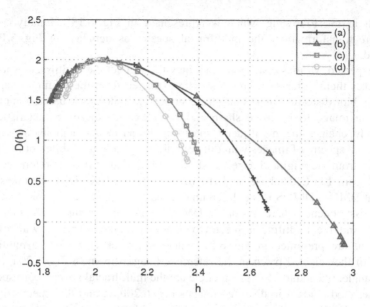

Fig. 6.21 Multifractal spectra of different distorted versions for "bikes"

quality. Image shown in Fig. 6.20d has lost much detail information, so the multifractal spectra of selected patch are most different from those of the reference image,
which suggests Fig. 6.20d has the worst image quality. These are all in accordance
with DMOS obtained by psychometric tests. Therefore, it is reasonable that such
discrepancies can be taken as features to assess the quality of distorted images.

6.5.3 Image Quality Degradation Quantification

Based on fractal description, the image quality-aware features take the form of
multifractal spectrum and fractal dimension. Distortions often change these features
of a test image in ways that make them dissimilar to that of an undistorted natural
image (an original image). Consequently, if we can quantify the discrepancies of
multifractal spectrum and fractal dimension between the test image and the original/
reference image, the visual quality degradation can be estimated accurately.

The discrepancies of multifractal spectra can be quantified by Euclidean distance
between each corresponding patch of the reference and distorted images. The
distance of all the moment order q is defined as,

$$L_i = \frac{\sum\limits_{q=-N}^{N} \sqrt{[D_{\mathrm{ref}}(h) - D_{\mathrm{dis}}(h)]^2 + (h_{\mathrm{ref}} - h_{\mathrm{dis}})^2}}{2N+1}. \qquad (6.104)$$

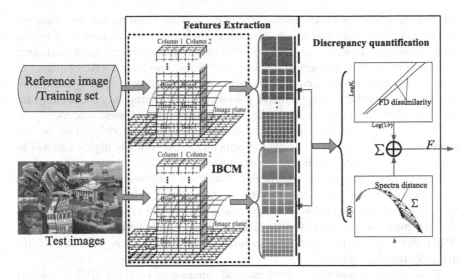

Fig. 6.22 Framework of image quality metric based on multifractal analysis

where N is the number of moment order q which is used to gain partition function. N should be neither too small to get the accurate multifractal spectra of an image nor too large to avoid overflow during calculation.

At last, the final distortion index is estimated as,

$$F = \frac{1}{M} \sum_{i=1}^{M} w_i \times [L_i + o(\Delta FD_i)] \qquad (6.105)$$

where M is the total number of patches that images are divided into, w_i are the weights of different patches to imitate that HVS is more sensitive to the irregular and erratic parts of images, $o(\cdot)$ is a function of ΔFD_i as a fine-tuning factor to get more accurate visual quality prediction, and ΔFD_i presents the dissimilarity of fractal dimension between the test image and the reference one.

The framework of image quality metric based on multifractal analysis is illustrated in Fig. 6.22. It should be noted that in the case of no-reference image quality assessment, since there is no any information of the original image, it is necessary to construct an ideal model for "reference." Such ideal model can be trained by a set of undistorted images. Consequently, deviation between actual features extracted from distorted images and the model reflects image quality degradation.

6.6   Summary

There are many metrics in IQA based on building systems to simulate HVS. Since the ultimate receptor of images is human beings, these metrics are meaningful to capture the features consistent with perceived information. However, the underlying mechanism of HVS is so complicated that it has not been well studied yet, there are still many functions of some cortex cells remain undiscovered and many mechanisms undetermined. In addition, the more complicated structures we discovered, the higher computational complexity it would cost, which would also limit the development of these methods. In (Wang and Bovik 2006), the four problems (quality definition, suprathreshold problem, natural image complexity, and dependency decoupling) are put forward to further clarify the limitations and difficulties of these models.

In this case, researchers attempt to construct the IQA models based on the properties of images. The fundamental principle is that HVS is highly adapted to extract statistical information from the visual scene. And as some distortions introduced in an image, the statistical characteristics are inevitably affected. The differences between statistical features extracted from reference and distortion images provide the measurement. IQA metrics based on natural image statistics achieve great success (Sheikh et al. 2005; Moorthy and Bovik 2011; Saad et al. 2012). Actually, the information from natural image statistics is not sufficient for image quality assessment.

Thus, many state-of-the-art IQA methods attempt to combine HVS properties and natural image statistics (Zhang et al. 2013a; Yan et al. 2016; Deng et al. 2017). In (Zhao et al. 2016), the authors introduce a combination approach by employing log Gabor filters to model the perceptual characteristics of visual cortex and local derivative pattern (LDP) to describe detailed texture information whose framework is shown in Fig. 6.23. With high accuracy and computational complexity, the metric has better performance than most of the existing IQA algorithms. Actually,

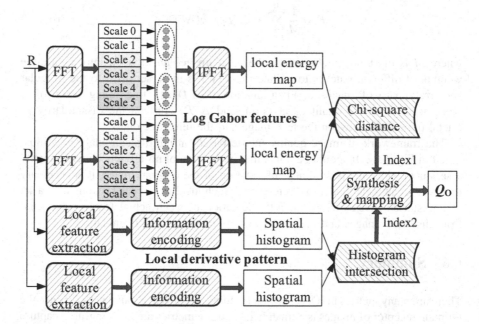

Fig. 6.23 Example of combination approach

properties of HVS can be employed as the assistant to natural image statistics-based metrics. One of the typical examples is SSIM. SSIM is firstly calculated in local regions, and the authors propose that the suitable weight of different regions can improve the metric's performance. The experiments in (Ding et al. 2012) prove saliency-based pooling is helpful to improve the accuracy of SSIM algorithm. In consequence, in order to construct a good IQA method satisfying three criteria including high accuracy, strong robust across different databases and distortion types and low computational complexity, these IQA metrics are supposed to combine HVS properties with natural image statistics.

References

Ahonen, T., Hadid, A., & Pietikainen, M. (2006). Face description with local binary patterns: Application to face recognition. *IEEE Transactions on Pattern Analysis and Machine Intelligence, 28*(12), 2037–2041.

Allain, C., & Cloitre, M. (1991). Characterizing the lacunarity of random and deterministic fractal sets. *Physical Review A, 44*(6), 3552–3558.

Asvestas, P., Matsopoulos, G. K., & Nikita, K. S. (1998). A power differentiation method of fractal dimension estimation for 2-D signals. *Journal of Visual Communication and Image Representation, 9*(4), 392–400.

Balghonaim, A. S., & Keller, J. M. (1998). A maximum likelihood estimate for two-variable fractal surface. *IEEE Transactions on Image Processing, 7*(12), 1746–1753.

Chang, H. W., Zhang, Q. W., Wu, Q. G., & Gan, Y. (2015). Perceptual image quality assessment by independent feature detector. *Neurocomputing, 151*(3), 1142–1152.

Chaudhuri, B. B., & Sarker, N. (2002). Texture segmentation using fractal dimension. *IEEE Transactions on Pattern Analysis and Machine Intelligence, 17*(1), 72–77.

Costa, M. F., Barboni, M. T. S., & Ventura, D. F. (2011). Psychophysical measurements of luminance and chromatic spatial and temporal contrast sensitivity in duchenne muscular dystrophy. *Psychology & Neuroscience, 4*(1), 67–74.

Deng, R., Zhao, Y. & Ding, Y. (2017). Hierarchical feature extraction assisted with visual saliency for image quality assessment. *Journal of Engineering,* 4752378.

Ding, Y., Zhang, Y., Zhang, D., & Wang, X. (2012). Weighted multi-scale structural similarity for image quality assessment with saliency-based pooling strategy. *International Journal of Digital Content Technology and its Applications, 6*(5), 67–78.

Ding, Y., Dai, H., & Wang, S. (2014). Image quality assessment scheme with topographic independent components analysis for sparse feature extraction. *Electronics Letters, 50*(7), 509–510.

Ding, Y., Zhang, H., Luo, X. H., & Dai, H. (2015). Blind image quality assessment based on fractal description of natural scenes. *Electronics Letters, 51*(4), 338–339.

Ding, Y., Chen, H. D., Zhao, Y., & Zhu, Y. F. (2016a). No-reference image quality assessment based on Gabor filters and nonlinear feature extraction. *International Journal of Digital Content Technology and its Applications, 10*(5), 100–109.

Ding, Y., Li, N., Zhao, Y., & Huang, K. (2016b). Image quality assessment method based on non-linear feature extraction in kernel space. *Frontiers of Information Technology & Electronic Engineering, 17*(10), 1008–1017.

Ding, Y., Zhao, X. Y., Zhang, Z., & Dai, H. (2017). Image quality assessment based on multi-order local features description, modeling and quantification. *IEICE Transactions on Information and Systems, E, 100D*(6), 1303–1315.

Du, S., Yan, Y., & Ma, Y. (2016). Blind image quality assessment with the histogram sequences of high-order local derivative patterns. *Digital Signal Processing, 55*, 1–12.

Engelke, U., Nguyen, V. X., & Zepernick, H.-J. (2008). Regional attention to structural degradations for perceptual image quality metric design. In *2008 IEEE International Conference on Acoustics, Speech and Signal Processing*, 869–872.

Eskicioglu, A. M., & Fisher, P. S. (1995). Image quality measures and their performance. *IEEE Transactions on Communications, 43*(12), 2959–2965.

Fan, K. C., & Hung, T. Y. (2014). A novel local pattern descriptor local vector pattern in high-order derivative space for face recognition. *IEEE Transactions on Image Processing, 23* (7), 2877–2891.

Field, D. J. (1987). Relations between the statistics of natural images and the response properties of cortical cells. *Journal of the Optical Society of America A, 4*(12), 2379–2394.

Fränti, P. (1998). Blockwise distortion measure for statistical and structural errors in digital images. *Signal Processing: Image Communication, 13*(2), 89–98.

Geng, X., Shen, L., Li, K., & An, P. (2016). A stereoscopic image quality assessment model based on independent component analysis and binocular fusion property. *Signal Processing Image Communication, 2017*(52), 54–63.

Ghosh, K., Sarkar, S., & Bhaumik, K. (2007). Understanding image structure from a new multi-scale representation of higher order derivative filters. *Image and Vision Computing, 25* (8), 1228–1238.

Goodman, J. S., & Pearson, D. E. (1979). Multidimensional scaling of multiply-impaired television pictures. *IEEE Transactions on Systems, Man, and Cybernetics, 9*(6), 353–356.

Gu, K., Zhou, J., Qiao, J.-F., Zhai, G., Lin, W., & Bovik, A. C. (2017). No-reference quality assessment of screen content pictures. *IEEE Transactions on Image Processing, 26*(8), 4005–4018.

Guo, Z., Zhang, L., & Zhang, D. (2010). A completed modeling of local binary pattern operator for texture classification. *IEEE Transactions on Image Processing, 19*(6), 1657–1663.

Guo, W., & Zong, Q. (2012). A blind separation method of instantaneous speech signal via independent components analysis. In *International Conference on Consumer Electronics*, 3001–3004.

Han, X.-H., Chen, Y.-W., & Gang, X. (2015). High-order statistics of weber local descriptors for image representation. *IEEE Transactions on Cybernetics, 45*(6), 1180–1193.

Henriksson, L., Hyvärinen, A., & Vanni, S. (2009). Representation of cross frequency spatial phase relationships in human visual cortex. *Journal of Neuroscience, 29*(45), 14342–14351.

Huang, L., Cui, X., Lin, J., & Shi, Z. (2011). A new reduced-reference image quality assessment method based on SSIM. *Applied Mechanics and Materials, 55*, 31–36.

Huang, D., Zhu, C., Wang, Y., & Chen, L. (2014). HSOG: A novel local image descriptor based on histograms of the second-order gradients. *IEEE Transactions on Image Processing, 23*(11), 4680–4695.

Humeau, A., Buard, B., Mahé, G., Chapeau-Blondeau, F., Rousseau, D., & Abraham, P. (2010). Multifractal analysis of heart rate variability and laser doppler flowmetry fluctuations: comparison of results from different numerical methods. *Physics in Medicine & Biology, 55* (20), 6279–6297.

Hyvärinen, A., Hoyer, P. O., & Inki, M. (2001). Topographic independent component analysis. *Neural Computation, 13*(7), 1527–1558.

Hyvärinen, A., Hurri, J., & Hoyer, P. O. (2009). *Natural image statistics: A probabilistic approach to early computational vision*. London: Springer-Verlag.

Ida, T., & Sambonsugi, Y. (1998). Image segmentation and contour detection using fractal coding. *IEEE Transactions on Circuits System Video Technology, 8*(8), 968–977.

Jain, R., Kasturi, R., & Schunck, B. G. (1995). *Machine Vision*. New York: McGraw-Hill.

Jiao, S., Qi, H., Lin, W., & Shen, W. (2013). Fast and efficient blind image quality index in spatial domain. *Electronic Letters, 49*(18), 1137–1138.

Jähne, B., Haubecker, H., & Geibler, P. (1999). *Handbook of computer vision and applications.* New York: Academic.

Kasturiwate, H. P., & Deshmukh, C. N. (2009). Quality assessment of ICA for ECG signal analysis. In *International Conference on Emerging Trends in Engineering and Technology,* 73–75.

Kovesi, P. (1999). Image features from phase congruency. *Journal of Computer Vision Research, 1*(3), 1–26.

Kruger, N., Janssen, P., Kalkan, S., Lappe, M., Leonardis, A., Piater, J., et al. (2013). Deep hierarchies in the primate visual cortex: What can we learn for computer vision? *IEEE Transactions on Pattern Analysis and Machine Intelligence, 35*(8), 1847–1871.

Larson, E. C., & Chandler, D. M. (2008). Unveiling relationships between regions of interest and image fidelity metrics. Proceeding of SPIE 6822, Visual Communications and Image Processing 2008, 68222A.

Larson, E. C., Vu, C. T., & Chandler, D. M. (2008). Can visual fixation patterns improve image fidelity assessment? 15th IEEE International Conference on Image Processing, 3: 2572–2575.

Lei, Z., Liao, S., Pietikäinen, M., & Li, S. Z. (2011). Face recognition by exploring information jointly in space, scale and orientation. *IEEE Transactions on Image Processing, 20*(1), 247–256.

Li, C. F., & Bovik, A. C. (2009). Three-component weighted structural similarity index. Proceedings of SPIE, 7242, image quality and system performance VI: 72420Q.

Li, J., Du, Q., & Sun, C. (2009). An improved box-counting method for image fractal dimension estimation. *Pattern Recognition, 42*(11), 2460–2469.

Li, J., Duan, L. Y., Chen, X., Huang, T., & Tian, Y. (2015). Finding the secret of image saliency in the frequency domain. *IEEE Transactions Pattern Analysis and Machine Intelligence, 37*(12), 2428–2440.

Liu, S., & Chang, S. (1997). Dimension estimation of discrete-time fractional Brownian motion with applications to image texture classification. *IEEE Transactions on Image Processing, 6* (8), 1176–1184.

Lin, K. H., Lam, K. M., & Siu, W. C. (2001). Locating the eye in human face images using fractal dimensions. *IEEE Proceedings on Vision, Image and Signal Processing, 148*(6), 413–421.

Liu, C., & Yang, J. (2009). ICA color space for pattern recognition. *IEEE Transactions on Neural Networks, 20*(2), 248–257.

Liu, D., Sun, D. M., & Qiu, Z. D. (2010). Feature selection for fusion of speaker verification via Maximum Kullback-Leibler distance. Signal Processing (ICSP), 2010 IEEE 10th International Conference on, Beijing, 565–568.

Loh, N., Hampton, C., Martin, A., Starodub, D., Sierra, R., Barty, A., et al. (2012). Fractal morphology, imaging and mass spectrometry of single aerosol particles in flight. *Nature, 486* (7404), 513–517.

Luo, Y. T., Zhao, L. Y., Zhang, B., Jia, W., Xue, F., Lu, J. T., et al. (2016). Local line directional pattern for palmprint recognition. *Pattern Recognition, 50,* 26–44.

Mancas-Thillou, C., & Gosselin, B. (2006). Character segmentation-by-recognition using log-Gabor filters. 18th International Conference on Pattern Recognition, 901–904.

Mandelbrot, B. B. (1967). How long is the coast of Britain? *Statistical self-similarity and fractional dimension. Science, 156*(3775), 636–638.

Mandelbrot, B. B., & Wheeler, J. A. (1983). The fractal geometry of nature. *Journal of the Royal Statistical Society, 147*(4), 468.

Marr, D., & Hildreth, E. (1980). Theory of edge detection. Proceedings of the Royal Society of London. *Series B, Biological Sciences, 207*(1167), 187–217.

Mendi, E. (2015). Image quality assessment metrics combining structural similarity and image fidelity with visual attention. *Journal of Intelligent & Fuzzy Systems, 28*(3), 1039–1046.

Meyer-Bäse, A., Auer, D., & Wismueller, A. (2003). Topographic independent component analysis for fMRI signal detection. *Proceedings of the International Joint Conference on Neural Networks, 1*(7), 601–605.

Mirny, L. A. (2011). The fractal globule as a model of chromatin architecture in the cell. *Chromosome Research, 19*(1), 37–51.

Mittal, A., Moorthy, A. K., & Bovik, A. C. (2012). No-reference image quality assessment in the spatial domain. *IEEE Transactions on Image Processing, 21*(12), 4695–4708.

Miyabe, S., Juang, B. H., Saruwatari, H., & Shikano, K. (2009). Kernel-based nonlinear independent component analysis for underdetermined blind source separation. In *IEEE International Conference on Acoustics,* 1641–1644.

Moorthy, A. K., & Bovik, A. C. (2009). Visual importance pooling for image quality assessment. *IEEE Journal of Selected Topics in Signal Processing, 3*(2), 193–201.

Moorthy, A. K., & Bovik, A. C. (2011). Blind image quality assessment: From natural scene statistics to perceptual quality. *IEEE Transactions on Image Processing, 20*(12), 3350–3364.

Morrone, M. C., Ross, J., Burr, D. C., & Owens, R. (1986). Mach bands are phase dependent. *Nature, 324*(6049), 250–253.

Morrone, M. C., & Burr, D. C. (1988). Feature detection in human vision: A phase-dependent energy model. Proceedings of the Royal Society of London. *Series B, Biological Sciences, 235* (1280), 221–245.

Murala, S., Maheshwari, R. P., & Balasubramanian, R. (2012). Local tetra patterns: A new feature descriptor for content-based image retrieval. *IEEE Transactions on Image Processing, 21*(5), 2874–2886.

Najemnik, J., & Geisler, W. S. (2005). Optimal eye movement strategies in visual search. *Nature, 434,* 387–391.

Neil, G., & Curtis, K. M. (1997). Shape recognition using fractal dimension. *Pattern Recognition, 30*(12), 1957–1969.

Nielsen, F., Hyvrinen, A., Hurri, J., & Hoyer, P. O. (2009). *Natural image statistics: A probabilistic approach to early computational vision.* New York: Springer-Verlag.

Ojala, T., Pietikäinen, M., & Harwood, D. (1996). A comparative study of texture measures with classification based on feature distributions. *Pattern Recognition, 29*(1), 51–59.

Ojala, T., Pietikainen, M., & Maenpaa, T. (2002). Multiresolution gray-scale and rotation invariant texture classification with local binary patterns. *IEEE Transactions on Pattern Analysis and Machine Intelligence, 24*(7), 971–987.

Olshausen, B. A., & Field, D. J. (1996). Emergence of simple-cell receptive field properties by learning a sparse code for natural images. *Nature, 381*(6583), 607–609.

Oszust, M. (2016). Full-reference image quality assessment with linear combination of genetically selected quality measures. *PLoS ONE, 11*(6), e0158333.

Peitgen, H. O., Jürgens, H., & Saupe, D. (2004). Chaos and fractals: New frontiers of science. *Mathematical Gazette, 79*(484), 241–255.

Pentland, A. P. (1984). Fractal-based description of natural scenes. *IEEE Transactions on Pattern Analysis and Machine Intelligence, 6*(6), 661–674.

Privitera, C. M., & Stark, L. W. (2000). Algorithms for defining visual regions-of-interest: comparison with eye fixations. *IEEE Transactions on Pattern Analysis and Machine Intelligence, 22*(9), 970–982.

Provata, A., & Katsaloulis, P. (2010). Hierarchical multifractal representation of symbolic sequences and application to human chromosomes. *Physical Review E: Statistical, Nonlinear, and Soft Matter Physics, 81*(2 Pt 2), 026102.

Rajashekar, U., Cormack, L. K., & Bovik, A. C. (2003). Image features that draw fixations. Proceedings 2003 International Conference on Image Processing, Barcelona, Spain.

Russell, D. A., Hanson, J. D., & Ott, E. (1980). Dimension of strange attractors. *Physical Review Letters, 45*(14), 1175–1178.

Saad, M., Bovik, A. C., & Charrier, C. (2012). Blind image quality assessment: a natural scene statistics approach in the DCT domain. *IEEE Transactions on Image Processing, 21*(8), 3339–3352.

Sampat, M. P., Wang, Z., Gupta, S., Bovik, A. C., & Markey, M. K. (2009). Complex wavelet structural similarity: A new image similarity index. *IEEE Transactions on Image Processing, 18*(11), 2385–2401.

Sarker, N., & Chaudhuri, B. B. (1994). An efficient differential box-counting approach to compute fractal dimension of image. *IEEE Transactions on Systems Man and Cybernetics, 24*(1), 115–120.

Sheikh, H. R., & Bovik, A. C. (2006). Image information and visual quality. *IEEE Transactions on Image Processing, 15*(2), 430–444.

Sheikh, H. R., Bovik, A. C., & Cormack, L. (2005). No-reference quality assessment using natural scene statistics: JPEG2000. *IEEE Transactions on Image Processing, 14*(11), 1918–1927.

Simoncelli, E. P. (1997). Statistical models for images: compression, restoration and synthesis. *Conference Record of the Asilomar Conference on Signals, Systems and Computers, 1,* 673–678.

Simoncelli, E. P., & Olshausen, B. A. (2001). Natural image statistics and neural representation. *Annual Review of Neuroscience, 24*(1), 1193–1216.

Srivastava, A., Lee, A. B., Simoncelli, E. P., & Zhu, S. C. (2003). On advances in statistical modeling of natural images. *Journal of mathematical imaging and vision, 18*(1), 17–33.

Stosic, T., & Stosic, B. D. (2006). Multifractal analysis of human retinal vessels. *IEEE Transactions on Medical Imaging, 25*(8), 1101–1107.

Tan, X., & Triggs, B. (2010). Enhanced local texture feature sets for face recognition under difficult lighting conditions. *IEEE Transactions on Image Processing, 19*(6), 1635–1650.

Wainwright, M. J., & Simoncelli, E. P. (1999). Scale mixtures of Gaussians and the statistics of natural images. *Gayana, 68*(2), 609–610.

Wang, Z. (2001). Rate scalable Foveated image and video communications. Ph.D. Dissertation, Department of Electrical and Computer Engineering, University. Texas at Austin, Austin, TX.

Wang, Z., & Bovik, A. C. (2002). A universal image quality index. *IEEE Signal Processing Letters, 9*(3), 81–84.

Wang, Z., & Bovik, A. C. (2006). Modern image quality assessment. *Synthesis Lectures on Image, Video, and Multimedia Processing, 2*(1), 1–156.

Wang, Z., Bovik, A. C., Sheikh, H. R., & Simoncelli, E. P. (2004). Image quality assessment: From error visibility to structural similarity. *IEEE Transactions on Image Processing, 13*(4), 600–612.

Wang, L., & He, D.-C. (1990). Texture classification using texture spectrum. *Pattern Recognition, 23*(8), 905–910.

Wang, Z., & Li, Q. (2011). Information content weighting for perceptual image quality assessment. *IEEE Transactions on Image Processing, 20*(5), 1185–1198.

Wang, W., Li, J., Huang, F., & Feng, H. (2008). Design and implementation of log-Gabor filter in fingerprint image enhancement. *Pattern Recognition Letters, 29*(3), 301–308.

Wang, Z., & Simoncelli, E. P. (2005). Reduced-reference image quality assessment using a wavelet-domain natural image statistic model. *Proceedings of the SPIE, 5666,* 149–159.

Wang, Z., Simoncelli, E. P., & Bovik, A. C. (2003). Multi-scale structural similarity for image quality assessment. *Conference Record of the Asilomar Conference on Signals, Systems and Computers, 2,* 1398–1402.

Wei, X. & Li, C. (2010). Visual saliency detection based on topographic independent component analysis. Signal Processing (ICSP), 2010 IEEE 10th International Conference on, Beijing, 1244–1247.

Wu, Q., Li, H., Meng, F., Ngan, K. N., Luo, B., Huang, C., et al. (2016). Blind image quality assessment based on multichannel feature fusion and label transfer. *IEEE Transactions on Circuits and Systems for Video Technology, 26*(3), 425–440.

Yan, Y. P., Du, S. L., Zhang, H. J., & Ma, Y. D. (2016). When spatial distribution unites with spatial contrast: An effective blind image quality assessment model. *IET Image Processing, 10* (12), 1017–1028.

Yu, L., Zhang, D., Wang, K., & Yang, W. (2005). Coarse iris classification using box-counting to estimate fractal dimensions. *Pattern Recognition, 38*(11), 1791–1798.

Yu, X., Hu, D., & Xu, J. (2014). Kernel independent component analysis. In *Blind source separation: Theory and applications* (pp. 145–152). Singapore: John Wiley and Sons.

Yuan, F. N. (2014). Rotation and scale invariant local binary pattern based on high order directional derivatives for texture classification. *Digital Signal Processing, 26*(1), 142–152.

Yuan, J., Wang, D., & Cheriyadat, A. M. (2015). Factorization-based texture segmentation. *IEEE Transactions on Image Processing, 24*(11), 3488–3497.

Zhang, Y., & Chandler, D. M. (2013). No-reference image quality assessment based on log-derivative statistics of natural scenes. *Journal of Electronic Imaging, 22*(4), 451–459.

Zhang, H., Ding, Y., Wu, P. W., Bai, X. T., & Huang, K. (2014). Image quality assessment by quantifying discrepancies of multifractal spectrums. *IEICE Transactions on Information and Systems, E, 97D*(9), 2453–2460.

Zhang, D., Ding, Y., & Zheng, N. (2012). Nature science statistics approach based on ICA for no-reference image quality assessment. *Procedia Engineering, 29*(4), 3589–3593.

Zhang, H., Gao, W., Chen, X., & Zhao, D. (2005). Learning informative features for spatial histogram-based object detection. *Proceedings IEEE International Joint Conference on Neural Networks, 3*, 1806–1811.

Zhang, B., Gao, Y., Zhao, S., & Liu, J. (2010). Local derivative pattern versus local binary pattern: Face recognition with high-order local pattern descriptor. *IEEE Transactions on Image Processing, 19*(2), 533–544.

Zhang, M., Mou, X., Fujita, H., Zhang, L., Zhou, X., & Xue, W. (2013a). Local binary pattern statistics feature for reduced reference image quality assessment. *Proceedings of SPIE, 8660* (3), 872–886.

Zhang, M., Muramatsu, C., Zhou, X., Hara, T., & Fujita, H. (2015). Blind image quality assessment using the joint statistics of generalized local binary pattern. *IEEE Signal Processing Letters, 22*(2), 207–210.

Zhang, F., & Roysam, B. (2016). Blind quality metric for multidistortion images based on cartoon and texture decomposition. *IEEE Signal Processing Letters, 23*(9), 1265–1269.

Zhang, L., Zhang, L., Mou, X., & Zhang, D. (2011). FSIM: A feature similarity index for image quality assessment. *IEEE Transactions on Image Processing, 20*(8), 2378–2386.

Zhang, L., Tong M. H., Marks, T. K., Shan, H., & Cottrell, G. W. (2008). SUN: A Bayesian framework for saliency using natural statistics. *Journal of Vision, 8*(7):32, 1–20.

Zhang, M., Xie, J., Zhou, X., & Fujita, H. (2013b). No reference image quality assessment based on local binary pattern statistics. 2013 Visual Communications and Image Processing (VCIP), Kuching, 1–6.

Zhao, Y., Ding, Y., & Zhao, X. Y. (2016). Image quality assessment based on complementary local feature extraction and quantification. *Electronics Letters, 52*(22), 1849–1851.

Zhou, W. X. (2008). Multifractal detrended cross-correlation analysis for two nonstationary signals. *Physical Review E: Statistical, Nonlinear, and Soft Matter Physics, 77*(6), 066211.

Zhou, W. J., Yu, L., Qiu, W. W., Zhou, Y., & Wu, M. W. (2017). Local gradient patterns (LGP): An effective local-statistical-feature extraction scheme for no-reference image quality assessment. *Information Sciences, 397*, 1–14.

Chapter 7
Stereoscopic Image Quality Assessment

7.1 Introduction

By far we have been restraining our topic within the field of planar (2D) image quality assessment (IQA), where the objects whose quality is evaluated are individual digital images. Playing a substantially important role for carrying visual signals in the last several decades, the 2D images are currently facing the challenge that stereoscopic (3D) images are gradually taking over many of their applications, which can be observed from the overwhelming demands from entertainment industry, e.g., 3D films, 3D cameras (which are more and more popularly equipped on handy devices such as smartphones), at the very least. With the development of 3D image processing techniques, especially due to the easy access to 3D scene acquisition and display facilities, knowledge about human binocular vision would be necessary for both academical and industrial purposes, certainly including study upon stereoscopic image quality assessment (SIQA).

The term stereoscopic image we refer to, or 3D image, stereo image pair, denotes a pair of images containing the same contents filmed from slightly different horizontal position that creates an illusion of depth information. Estimating depth is one of the most important responsibility that human binocular vision bears. Because objects are viewed by two eyes from different positions and angles, the visual cortex can intuitively analyze visual signals from both eyes and construct 3D vision. Therefore, simply shooting an object with two cameras placed at the same horizontal level is sufficient for modeling inputs of visual signal; similarly, displaying the different visual signals generated by the two-camera system for both eyes helps the visual cortex to develop 3D vision. The two images that constitute a stereoscopic image are commonly referred to as left view and right view.

Figure 7.1 presents a stereoscopic image from LIVE 3D IQA database (Moorthy et al. 2013). The locations that objects appear in the left view (a) are all in the right of that in the right view (b). For instance, in the left and right views, we denote d_1 and d_2 as the distance from the left border of the view to the base of the statue,

© Zhejiang University Press, Hangzhou and Springer-Verlag GmbH Germany 2018   161
Y. Ding, *Visual Quality Assessment for Natural and Medical Image*,
https://doi.org/10.1007/978-3-662-56497-4_7

(a)   (b)

Fig. 7.1 **a** Left view and **b** right view of a stereoscopic image

respectively. Obviously, we have $d_1 > d_2$. And each object is on the same horizontal level for both views, of course.

However, the seemingly straightforward acquisition and display process can lead to complicated problems, some of which are yet to be understood. In short, human visual system (HVS) does not take the mix of binocular visual signals as simple summation, but constructs 3D scenes through a very complex process. The fact that the constructing process from two separate views to a mental 3D scene is more than simple summation has been confirmed by studies in subjective SIQA. For one famous instance, experiments in (Meegan et al. 2001) discovered that when viewing stereoscopic images with the views at different quality level, the perceived quality usually tends to be close to that of either view rather than maintains as the average of the views, and this tendency was highly distortion-dependent. It was assumed, and is now generally accepted, that the view with more information (or higher energy, another way to put it) would receive more attention and has more power in defining the quality of the stereo image pair. Such findings confirm the existence of sophisticated HVS properties affecting stereoscopic image quality perception. An illustration is given in Fig. 7.2, where the right view of the stereoscopic image in Fig. 7.1 is distorted by (a) JPEG compression and (b) Gaussian blur. According to (Meegan et al. 2001), when only one view is distorted, if it is a block-based compression like (a), the perceived quality would be as worse as the distorted view; while if the view is blurred like (b), the perceived quality would be close to the higher-quality view as if the distortion is absent. This is because the blur deduces the energy in images while the block edges caused by JPEG compression increases the energy.

From another aspect, it is considered that the perceived quality of stereoscopic images is a "generalized" concept that perceived depth, sharpness, eyestrain, etc., should all be taken into account aside from the respective quality of the two 2D views (Seuntiëns et al. 2006). Kaptein et al. (2008) conclude the factors apart from traditional quality definition as the "naturalness" of stereoscopic images and urge researches upon it, especially upon its relation to depth information and its contribution to 3D visual experience. Although it is not quite clear that on what degree and in what way the perceived stereoscopic image quality is influenced, it is certain

(a)

(b)

Fig. 7.2 A stereoscopic image with right view distorted by **a** JPEG compression and **b** Gaussian blur

that the perceived naturalness, or visual comfort, is affecting subjective SIQA results.

The evidence for the fact that binocular vision is not simple signal summation in objective SIQA is that directly applying 2D IQA methods for SIQA, a common effort in early works (Benoit et al. 2008b; Yasakethu et al. 2008; You et al. 2010), often fails to achieve satisfactory results. The unexpected influence from binocular vision generation aside from visual signal summation is summarized as binocular effects. Recent approaches that take the binocular effects into account using various means, as expected, are capable of achieving much better performance than the direct employment of 2D methods. Binocular effects are the main enemy SIQA researchers are confronting, yet they are exactly the existence of binocular effects that let their study make sense, since otherwise the 2D IQA methods would be sufficient in the 3D field. Human binocular vision, including some of the appealing binocular effects having influences upon stereoscopic image quality perception, will be discussed in Sect. 7.2.

Similar to 2D IQA, objective SIQA methods are categorized as full-reference, reduced-reference, and no-reference, and the evaluation of the performance of a method is based on how well it correlates with subjective judgement. Subjective SIQA and existing databases will be introduced in Sect. 7.3. Detailed discussions about current objective SIQA methods are provided in Sect. 7.4.

7.2   Binocular Vision

7.2.1   Binocular Fusion

Binocular vision is possessed by many creatures. For them, having two eyes instead of one brings about many advantages, among which two appealing ones are the obtaining of a wider field of view and the generation of depth perception (Fahle 1987). There exists an interesting trade-off: The depth perception occurs only in the overlapping viewing field of both eyes, so it is generated at the cost of a wide field of view. Usually but not always, herbivorous animals (those eat grass or vegetables as food) have a wide viewing angle but very narrow overlapping area to better detect danger in the surroundings, while predatory animals (those get protein mainly by eating meat) tend to have two eyes positioned on the front of their heads to gain a smaller viewing angle in exchange for better stereoscopic vision. Humans obviously belong to the latter category. Visions in the overlapping and nonoverlapping viewing areas of the eyes are called binocular vision and monocular vision, respectively, as shown in Fig. 7.3. Monocular vision, as you can imagine, only occurs either in the fields that are close to the left or right border of viewing field where can only be seen by one eye. For human, in most cases, objects could be seen by both of our eyes and binocular vision is obtained. When the binocular vision is generated, the two input visual signals are fused into one, with additional depth information contained. This phenomenon is called binocular fusion.

The single scene constructed in our mind is called a cyclopean image. In reality, humans do not always sense that they are using two eyes. Instead, they almost always feel they are seeing only one scene, that is, binocular fusion is almost always happening, and we even do not realize (Julesz 1971). The cyclopean images are different from ordinary 2D images for the depth information it contains, meaning the perceived distance between the viewer and objects. A fun fact is that the term "cyclopean image" is from the one-eyed giant Cyclops in Homer's Odyssey; however, we can be quite sure that since Cyclops cannot obtain binocular vision,

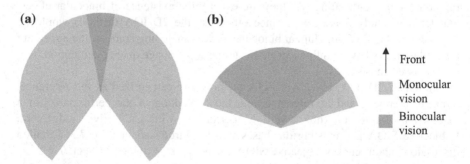

Fig. 7.3 A rough illustration of the viewing fields of typical **a** herbivorous animals and **b** humans

it is impossible for him to perceive cyclopean images. A disappointing fact is that, even with current knowledge about neuroscience, it is still not clear how the cyclopean image construction is really processed in our minds.

Benefits of binocular fusion can be concluded from at least two aspects. An immediate advantage is that the visual signals are added up by HVS, a process called binocular summation, and lead to the reinforcement of vision and enhance the ability to detect faint objects (Blake and Fox 1973). Although usually neglected, but we evident see things more clearly when using two eyes instead of one. Of course, the more important gain of having binocular vision is the generation of stereopsis, which means the different locations that a viewed object projects for left and right eyes are computed intuitively in the visual cortex and gives precise perception of depth, as will be discussed in the following.

7.2.2   Stereopsis

Stereopsis plays a key role in developing stereoscopic vision (Howard and Rogers 1995). As is previously mentioned, humans, along with many other animals (typical examples are other primates and the predators) that have a large binocular vision field and relatively small viewing angle, the two eyes are both facing the front with different lateral positions. An object in the binocular vision area projects on the different locations of the retinas of both eyes. This difference, by nature, can only be horizontal and is generally denoted as binocular disparity.

Evidently, for a viewed object, when it moves further from the viewer, the binocular disparity would become smaller and vice versa. To put it more formally, larger (or smaller) depth will be perceived by human due to the smaller (or larger) binocular disparity it causes. Figure 7.4 gives an illustration about this fact.

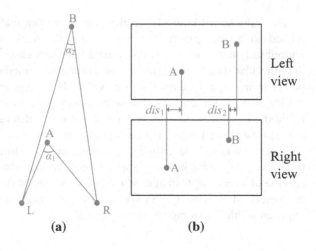

Fig. 7.4 Difference of object depths reflected in **a** different viewing angles of left and right eyes and **b** different position shifts in left and right views

Figure 7.4a is a vertical view of how human eyes (points L and R) perceive objects (simplified as points A and B). Evidently, for objects nearer to the viewer (A in this case), binocular disparity would be more significant, shown as a larger angle between the connecting lines of the object and both eyes (α_1). Reflected in the stereoscopic images as (b), the position shift of A in both views (dis$_1$) would be larger than that of B (dis$_2$).

7.2.3 Binocular Rivalry

Since a very long time ago, it has been discovered that the perceived quality of a stereoscopic image is related in a complicated way to its two views. As early as in 1960, an informal observation is put forward that lowering the resolution of a view has little affection on the apparent sharpness of the stereoscopic image (Julesz 1960). Based on this observation, the perceived quality of a stereoscopic image seems to be defined by the relatively higher-quality view, but later study shows some opposite results. Especially, experiments in (Meegan et al. 2001) found that the perceived stereoscopic image quality tends to be close to the higher-quality view for blurring, but tends to be close to the lower-quality view for blocky. Cause of this distortion-dependence was assumed by the researchers then that HVS is more interested in the view containing more information. It is now acknowledged that HVS indeed pays more attention to the view with more information, or to put it another way, with larger energy. When the quality difference of the left and right views is sufficient to be noticed, or sometimes described as when it is enough to result in a mismatch between the views, *binocular rivalry* occurs. When it does, the view with larger energy attracts more attention from the HVS, in other words it is in dominance. Note that when the images are locally distorted, i.e., the distortion is not uniformly distributed, the dominance might not be constant for different regions in an image.

The influence of binocular rivalry is so important that its introduction is closely related to image quality perception from the very beginning, and it is now undoubtedly the most widely considered binocular effect in modern SIQA study. If the binocular rivalry is absent, we can compute the quality of a stereoscopic image barely by averaging the quality of its views. How to properly account for binocular rivalry is a major obstacle but a necessary part in designing objective SIQA methods. In later sections, when we move on to detailed objective methods, the concept binocular rivalry will be brought up repeatedly.

It is worth noted that although it seems unwanted from some aspects, binocular rivalry can be useful in many applications. An important utility is a compression scheme of stereoscopic image: When one view goes through blur-like compression, the perceived stereoscopic image quality can maintain at a high level when the required width for storage is largely saved.

7.2.4   Ocular Dominance

Another important binocular effect is ocular dominance, also called eye preference, eyedness, denoting the HVS's preference to visual input from one eye to another (Khan and Crawford 2001). This concept, unlike binocular rivalry, concerns a born property rather than a content-based preference. According to previous studies (Reiss 1997; Ehrenstein et al. 2005; Eser et al. 2008), about two-thirds of the population is right-eye dominant and one-third left-eye dominance, and a small portion of people belongs to neither of the classes.

Also, ocular dominance can easily be confused with laterality. Laterality defines one's preference for one side of the body, including left/right-handiness and left/right-footedness. But the preference for eye and body does not always match, indicating the ocular dominance is not part of laterality (Porac and Coren 1975). The internal mechanisms are different, each of the two hemisphere controls one side of the body, but each hemisphere controls half of both retinas. There is no proof on the existence of the connection between the ocular dominance and laterality.

Ocular dominance poses a threat to SIQA because we are not sure whether the left-eye dominant population perceives stereoscopic image quality differently from the right-eye dominant. The fact that right-eye dominance is in majority among people is not persuasive enough for us to endowing more weights for the right view when operating SIQA. In practice, the influence caused by ocular dominance is neglected, due partly to the unpredictability of the viewers. But in most cases, we can safely assume that if there exists binocular rivalry, the influence of binocular rivalry would be much greater than that of the ocular dominance; while if the views perfectly match, giving different weights for them would be unnecessary. As a result, what we can do at the current stage is only to ignore ocular dominance, but the present outcome seems to be satisfactory even with this ignorance, although the ocular dominance is a very important issue when talking about binocular vision.

7.2.5   Visual Discomfort

We have briefly introduced the 3D scene acquisition and display techniques used in modern systems, showing the audiences two lateral different images. This scheme can indeed create pseudo-3D scenes, but is still inadequate for faking natural scenes. One thing it fails to take care of is that the eyes have to constantly focus on a screen, while in natural scenarios the focus of the eyes varies according to objects. The distance between the eyes and the viewed object is called focal distance. A fixed focal distance would lead to visual discomfort, reflecting in symptoms including eye strain, headache, fatigue, asthenopia, leading to unpleasant viewing experience (Lambooij et al. 2009). The degree of visual discomfort is mainly related to the depth information in the stereoscopic images. Commonly, statistical features like larger depth, wider depth range, ununiform depth distribution lead to more severe visual discomfort (Yano et al. 2002).

The relation between visual discomfort and perceived stereoscopic image quality has not been studied thoroughly. Most existing SIQA methods choose to neglect the influence of the former on the latter. One evidence for their connection is that the occurrence of visual discomfort leads to undesirable viewing experience, and so does the distortion. Therefore, it is possible that the causes of visual discomfort bring about similar viewing experience as the effects of distortions. We can also make reasonable assumption about the influence the perceived image quality has on visual discomfort. Current study upon visual discomfort is mainly based on undistorted images, because visual discomfort estimating methods would require a visual disparity approximation process to predict depth information. However, with distorted images, people might be incapable of obtaining a stereopsis and estimating the depth, so the level of visual discomfort must be influenced. To conclude, although not much related knowledge can be found in the literature, how strong visual discomfort and stereoscopic image quality affect each other demands studying.

7.3 Subjective Stereoscopic Image Quality Assessment

7.3.1 Principle

The way to evaluate the performance of SIQA methods is similar to that in the 2D field, as explained in Chaps. 2 and 4. In short, objective methods showing higher coherency with subjective results are considered better. Therefore, a crucial part in subjective SIQA study is building SIQA databases. The database construction is to some extent very alike as in 2D IQA, Chap. 2 gives a detailed description about how subjective experiments are conducted and how databases are constructed.

Therefore, the important issue here is the difference between the subjective study on 2D IQA and SIQA. We will discuss about only one thing: the introduction of asymmetrical distortion. Because a stereoscopic image is consisted of two views, the views are possibly distorted either symmetrically or asymmetrically. This possibility is unique for SIQA, comparing to 2D IQA, but it is not the most crucial reason why asymmetrical distortion is so important. We have mentioned more than once that binocular rivalry is making SIQA study meaningful. In most cases, only when assessing the quality of asymmetrically distorted stereoscopic images will binocular rivalry have to be considered, since the views mismatch much more often. Actually, it has been claimed that averaging the quality of both views gives proper prediction upon stereoscopic image quality (Yang et al. 2010). Thus, it is important to include asymmetrically distortion in subjective databases, and objective methods can only be proved good when it performs well on the asymmetrically distorted images. If an objective SIQA method performs good on symmetrically distorted images but fails to show similar results on asymmetrically distorted ones, it can only be validated that it is equipped with a good 2D IQA method.

7.3.2 Databases

In the following, we will list and discuss the known SIQA databases with the best of our knowledge. The databases will be introduced basically in the order of their publishing time, starting from the oldest.

(1) IRCCyN/IVC 3D Images Database (Benoit et al. 2008a)
   The builders of this database are from the team Image and Video-Communications (IVC), Institut de Recherche en Communication et Cybernétique de Nantes (IRCCyN), Nantes, France. The database contains 96 images symmetrically distorted by JPEG 2000 compression, JPEG compression, and Gaussian blur. Its evident disadvantages are that too few images are included and that the distortion is only symmetrical.

(2) MICT 3D Image Quality Evaluation Database (Sazzad et al. 2009)
   The builders are from Graduate School of Science and Engineering, University of Toyama, Japan. The database contains 500 distorted images, including both symmetrical and asymmetrical distortion, but JPEG compression is the only distortion type it considers.

(3) NBU 3D Image Quality Assessment Database Phase I & II (Wang et al. 2009; Zhou et al. 2011)
   The builders of the databases are with the Faculty of Information Science and Engineering, Ningbo University (NBU), Ningbo, China. Phase I of the database contains 410 images with asymmetrical distortion by JPEG 2000 compression, JPEG compression, white noise, and Gaussian blur, while Phase II contains 324 symmetrically distorted images, and H.264 compression is added for a new distortion type.

(4) TJU 3D Image Quality Assessment Database (Yang et al. 2009)
   This database also contains only symmetrically distorted images. The included 300 images suffer from JPEG 2000 compression, JPEG compression, and white noise. The builders are with the School of Electronic Information Engineering, Tianjin University (TJU), Tianjin, China.

(5) MMSPG 3D Image Quality Assessment Database (Goldmann et al. 2010)
   The builders are from Multimedia Signal Processing Group (MMSPG), Ecole Polytechnique Fédérale de Lausanne, Lausanne, Switzerland. This database is quite unique that the distortion is set as the different viewing distances of the images.

(6) LIVE 3D Image Quality Assessment Database Phase I & II (Moorthy et al. 2013; Chen et al. 2013b)
   The databases are constructed by the famous Laboratory for Image and Video Engineering (LIVE), University of Texas at Austin, Austin, United States.

Researchers with LIVE have contributed a lot for SIQA as well as 2D IQA. The two phases both contain images distorted by five distortion types: JPEG 2000 compression, JPEG compression, white noise, Gaussian blur, and fast-fading in Rayleigh Channel, same as in LIVE 2D IQA database. Phase I contains 365 symmetrically distorted images, and Phase II contains 360 images including both symmetrical and asymmetrical distortions.

(7) Waterloo-IVC 3D Image Quality Database Phase I & II (Wang et al. 2015)

The databases are built by researchers with the Department of Electrical and Computer Engineering, University of Waterloo, Waterloo, Canada. Three types of distortion are included, i.e., white noise, Gaussian blur, and JPEG compression. The two phases both contain images with either symmetrical or asymmetrical distortions while Phase II has more varying image contents. As claimed by the builders, there are two unique features about the databases. Firstly, they provide the subjective evaluation results for not only the stereo image pair, but also all single views they contain. Secondly, mixed image distortions are applied for some of the images.

For a more illustrative comparison, the basic information of the listed databases is summarized in Table 7.1.

7.4   Basic Frameworks for Objective SIQA

7.4.1   Estimating Depth by Computing Disparity

A straightforward idea to extend 2D IQA to SIQA is to introduce the analysis about the depth information (Benoit et al. 2008b). We are now aware that in a stereoscopic image, the depth of any object is always negatively related to the disparity of

Table 7.1 Basic information about the databases

	Year	Nationality	Symmetrical distortion contained	Asymmetrical distortion contained	Number of distortion types
IRCCyN/IVC	2008	France	Yes	No	3
MICT	2009	Japan	Yes	Yes	1
NBU-I	2009	China	No	Yes	4
NBU-II	2011	China	Yes	No	5
TJU	2009	China	Yes	No	3
MMSPG	2010	Switzerland	Yes	No	1
LIVE-I	2013	USA	Yes	No	5
LIVE-II	2013	USA	Yes	Yes	5
Waterloo-I	2015	Canada	Yes	Yes	3
Waterloo-II	2015	Canada	Yes	Yes	3

its appearance in the left and right views. Thus, the estimation or computation of depth and disparity is basically the same concept. We denote a matrix containing the disparity information or depth information of a stereoscopic images as its *disparity map* or *depth map*, respectively.

A disparity map or a depth map usually has to be based on either view. Without losing generality, we will always take left view as the basis in this book. Denote the disparity map and depth map as matrices **Di** and **De**, respectively, and denote the left and right views as **L** and **R**. Each element of a depth map, **De**(x, y), is a number representing the depth of the point **L**(x, y), i.e., how far the object presented by the point is from the cameras. As shown in Fig. 7.5, we denote x and y as vertical and horizontal coordinates, respectively, starting from the top-left corner. For **L**(x, y), let its corresponding point on the right view be **R**(x, y'), then the value of the corresponding point on the disparity map is

$$\mathbf{Di}(x, y) = y' - y \tag{7.1}$$

Obviously, the unit for **Di**(x, y) is the number of pixels (but not necessary integers). Although it is different from the unit for **De**(x, y), if we merely consider the absolute values, we can always observe the negative relation between **Di**(x, y) and **De**(x, y), which can be formally put as

$$[\mathbf{De}(x_1, y_1) - \mathbf{De}(x_2, y_2)][\mathbf{Di}(x_1, y_1) - \mathbf{Di}(x_2, y_2)] \leq 0 \tag{7.2}$$

where (x_1, y_1) and (x_2, y_2) denote two points. The left and right ends equal to each other only when **De**(x_1, y_1) = **De**(x_2, y_2).

However, two problems occur during the depth/disparity estimation. For one thing, it is not easy to precisely obtain the ground-truth depth data with the computed disparity, even assuming the computed disparity is ideally accurate. The projection from disparity to depth is complex and nonlinear, and involves trigonometric functions, not to mention the necessary input, the distance between the two cameras, is sometimes unknown. The common solution to this problem is to neglect the effects that the nonlinearity might bring and directly use the disparity maps to replace the functions of depth maps. Fortunately, this neglection shows no evident influence in most scenarios.

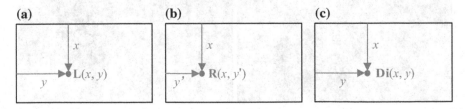

Fig. 7.5 A simplified illustration of what an element in a disparity map represents. **a** Left view and **b** right view of a stereoscopic image; **c** its disparity map based on the left view

For another, the assumption about the accuracy of the computed disparity is almost never ideally accurate, which is a much larger obstacle. Present disparity estimation schemes mainly follow the flow that for each pixel on the left view $L(x, y)$, its corresponding pixel on the right view $R(x', y')$ is required to be:

(1) On the same row: $x' = x$;
(2) To the left of (x, y): $y' > y$;
(3) Most similar to $L(x, y)$ among all candidates selected by (1) and (2).

The definition of "similar" in (3) can be interpreted in many ways. A most commonly adopted approach is to compare the sum of absolute value (SAD) of the neighbors of the pixel, usually based on a block matching strategy (Konolige 1997). To improve the accuracy, a semi-global block matching is presented in (Hirschmuller 2005) that additionally forces similar disparity on neighboring blocks. Another way for accuracy adjustment is to replace SAD. In (Chen et al. 2013b), the authors employed structural similarity (SSIM) (Wang et al. 2004) for this purpose and reported its superiority than SAD. However, there is no existing disparity estimation scheme that can be ideally accurate.

Figure 7.6 shows the depth map (c) and disparity map (d) of the stereoscopic image in Fig. 7.1. The disparity map is estimated using the scheme from (Chen et al. 2013b). The two above-mentioned problems are can be observed: For regions where the disparity is well estimated, the elements in corresponding locations in the depth and disparity maps are negative correlated; more importantly, there exist many regions where the disparity is ill computed.

Fig. 7.6 **a** Left view, **b** right view, **c** depth map, and **d** disparity map of a stereoscopic image

7.4.2 Analyzing Depth for SIQA

We have been talking about the unsatisfactory effectiveness of existing disparity estimation approaches, but what we did not mention is that the estimating results become even worse when the image is distorted. This fact, although undesirable, inspires a thought for researchers in early stages of objective SIQA study: Is the disparity map derived from a distorted stereoscopic image a good reflection of the distortion degree?

The objective full-reference method proposed in (Benoit et al. 2008b) is a good example that implements this thought. We choose this method as an example because of its representativeness, not because of its uniqueness. Many other works share same or similar framework with it.

The framework is shown in Fig. 7.7, which is very simple and clear. The notations "L." and "R." denote the left and right views, respectively. The framework aims to improve the traditional approach that directly averages the quality scores of both views to yield the final quality for stereoscopic images, by introducing an additional index, the difference between the two disparity maps computed using the distorted stereoscopic image and the reference.

The 2D IQA methods applied for both views are borrowed from what is proposed in (Campisi et al. 2007), which is also an early attempt in SIQA that we would like to briefly talk about before moving on to (Benoit et al. 2008b). In this work, the authors adopt a very simple two-stage flow. The quality scores of the two 2D images, left and right view, are first computed, respectively. Then, the respective quality scores are mapped to the final score for the stereo pair through various means.

Specifically, four 2D IQA metrics are selected for the first stage, including SSIM (Wang et al. 2004), Universal Quality Index (UQI) (Wang and Bovik 2002), C4 (Carnec et al. 2003), and Reduced Reference Image Quality Assessment (RRIQA)

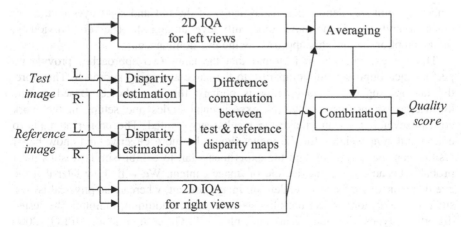

Fig. 7.7 Framework of the method proposed in (Benoit et al. 2008b)

(Wang and Simoncelli 2005). The first three are full-reference methods, and the last one is reduced-reference, so the evolved SIQA metric is accordingly full-reference and reduced-reference, basically.

In the second stage, three mapping strategies are tried to generate the final stereoscopic quality scores, which are referred to by the authors as the average approach, the main eye approach, and the visual acuity approach. These approaches determine how the weights are defined for the views. To state the problem clearly here, we have to do some spoiler. The authors actually adopt a weight-endowing-based SIQA approach, which is still one of the major attempts in designing modern methods that will be discussed in detail in the following sub-sections. Concretely, the average approach is to directly regard the quality of a stereo pair, Q, as the average of that of the left and right views, Q_L and Q_R,

$$Q = \frac{Q_L + Q_R}{2} \tag{7.3}$$

The main eye approach is based on the effect of ocular dominance, as introduced in Sect. 7.2, which suggests that there exists a preferred eye for most people. The quality is then defined as

$$Q = \frac{n_L Q_L + n_R Q_R}{n_{\text{total}}} \tag{7.4}$$

where n_L and n_R denote the number of invited observers for subjective quality evaluation that are left-eye preferred, right-eye preferred, and n_{total} is the total observer number equaling to the sum of n_L and n_R. The third approach, which is based on the visual acuity, is defined as

$$Q = \frac{1}{n_{\text{total}}} \left(\frac{\sum_i A_{L,i}}{A_{\max}} Q_L + \frac{\sum_i A_{R,i}}{A_{\max}} Q_R \right) \tag{7.5}$$

where $A_{L,i}$ and $A_{R,i}$ denote the visual acuity of the left and right eyes of the i-th observer, and $A_{\max}$ is the maximum acuity level among all observers. Obviously, the assumption behind this approach is the eye dominance.

Through experiments, it is found that the latter two approaches provide no performance improvement comparing to the naïve averaging approach. Therefore, the authors suggest the averaging is perhaps the best choice. Based on the knowledge we possess today, the experimental logics and setups in the work contain several flaws and make the results very limited. The proper attempts to endow different weights for the views should be more effective and show better results than the averaging, and the experiments fail to exhibit similar results most probably because of the neglection of image content. We will show later that the eye dominance is mostly dependent on image content, whereas the physical issues, such as ocular dominance, are almost irrelevant. In addition, although the tested distortion types include Gaussian blur, JPEG compression, JPEG 2000 compression, and the mixture of JPEG and JPEG 2000, the distortions are always symmetrically imposed on the left and right views. For symmetrically distorted stereoscopic images, their quality is indeed quite close to the average of the left and right views, while the logic does not hold for asymmetrical situations. Nevertheless, the method in (Benoit et al. 2008b) embraces the findings, putting an additional feature related to object depths with no other modification upon the 2D methods or the averaging, yet performance improvement is witnessed by all, so the fact that using disparity to estimate depth information is useful is validated.

The disparity estimation scheme adopted in (Benoit et al. 2008b) is the one presented in (Felzenszwalb and Huttenlocher 2006) and, as the authors claim, they have no intention to judge whether it is a good choice for disparity estimation or what scheme would be the best choice, and their aim is merely to discover whether the disparity maps would be helpful in SIQA. The distance between the disparity maps of the tested and reference images is computed by correlation coefficient (CC). CC is with the form

$$CC(x,y) = \frac{\sum_{i=1}^{N} (x_i - \mu(x))(y_i - \mu(y))}{\sqrt{\sum_{i=1}^{N} (x_i - \mu(x))^2 \sum_{i=1}^{N} (y_i - \mu(y))^2}} \tag{7.6}$$

where x and y are data with item number N, x_i is the i-th item in x, and μ denotes the averaging operation.

So far, we have both the 2D quality of the stereo pair derived as the average of the left and right views and the distance derived from the disparity maps. Let A and D denote them, respectively, and the left work is to find a way to synthesize them to the final stereoscopic image quality Q. The authors have come up with two ways to do so, represented as

$$Q = M \cdot \sqrt{D} \tag{7.7}$$

$$Q = M \cdot (1 + D) \tag{7.8}$$

In experiments, it turns out that computing Q by either of the two above equations outcomes better results than using M or D alone. Certainly, using more complex regression models is likely to develop better results, but the reported experimental results are sufficient to safely state the importance and effectiveness of the disparity information for SIQA.

Despite experimental results, the simplicity of this framework arouses doubts. The major concern is that the process of 2D IQA and that of depth analysis is separated, while this separation has no biological basis. HVS will not treat them separately, the evaluation of stereoscopic image quality in HVS is more likely through a serial process that combines the two images, or the responses of two images, or quality assessing results of two images, for further mapping to the final quality scores. In the next two sub-sections, we will present two kinds of methods that are more likely to be following this serial or cascaded approach. There exist other concerns as well. The reflection of distortion on the disparity map to stereoscopic image quality holds, to some degree, but maybe not sufficiently honest, and might dependent largely upon other issues such as the selection of disparity estimation scheme. For example, if a well-designed disparity estimation scheme is highly resistant to noise, then the distortions, even if very serious, might not reflect on the disparity maps, leading to an overestimate upon image quality. In addition, binocular effects such as binocular rivalry are totally neglected. We can safely make predictions that because the 2D scores are directly averaged, the method would perform especially poorly for asymmetrically distorted images.

Nevertheless, the role that disparity maps play for SIQA is still attached with great importance, of which the reason is at least twofold. On one hand, depth information is considered as an important property in stereoscopic perception. In the following sub-sections, we will introduce two main frameworks in modern SIQA. The first one is by constructing cyclopean images, the definition of which, as well as how they can be built, has been given previously. Through any known cyclopean image constructing scheme, estimating object depths by disparity maps is an evitable step. The second framework is by synthesizing the quality scores of the two planar images, the left and right views. This framework does not make use of the depth information explicitly, but this neglection attracts criticisms. So, methods adopting this flow usually construct disparity maps as well, to make complementary, to some extent (Xu et al. 2017). On the other hand, depth information is closely related to the degrees of visual discomfort of images and videos (Nojiri et al. 2003; Kim and Sohn 2011; Richardt et al. 2011; Choi et al. 2012; Park et al. 2014; Chen et al. 2017). The introduction of artificial depth information gives the stereoscopic images an inherent influence upon viewing experience. From this prospective, it is very reasonable that the disparity maps should be considered when evaluating perceptive quality of images.

7.4.3 Constructing Cyclopean Images for SIQA

As is discussed, object depth information is reasonably useful for evaluating the quality of stereo pairs. However, it is highly possible that separately extracting features from the 2D views and the disparity maps is not the best way to make use of the depth. Disparity maps ought to be connected to real image contents, and one choice to do so is by constructing cyclopean images. In this way, the information from the image content and depth are both preserved, and the subsequent procedures that conduct quality assessment for cyclopean images seem as simple as 2D IQA.

The full-reference method proposed in (Chen et al. 2013b) is a typical example adopting the cyclopean image constructing idea, of which the framework is given in Fig. 7.8. The framework can be virtually departed as three serial stages, disparity estimation, cyclopean image construction, and 2D IQA methods applied on the cyclopean images.

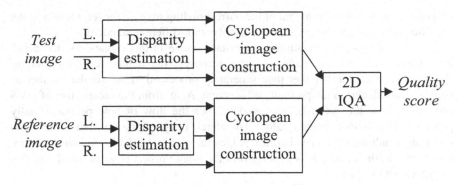

Fig. 7.8 Framework of the method proposed in (Chen et al. 2013b)

The first stage, disparity estimation, has actually been introduced in previous sections in this chapter. The difference between the topic we are discussing here and the previous introduction is that the original purpose of estimating disparity is to use disparity to model object depths (see Sect. 7.4.1), but here the disparity is directly employed in subsequent stage, i.e., cyclopean image construction. To recap, the disparity map **Di** is a matrix with the same size as the left or right views, in which each item represents how far the corresponding pixels in the left and right views are from each other. Concretely, assuming the two cameras for stereoscopic image acquisition are on the same horizontal level, an object would appear on the same horizontal position in the two views. Let the left view **L** be the baseline, for a pixel from it, $\mathbf{L}(x, y)$, denoting x as the vertical axis representing horizontal level, and y the horizontal axis, then its corresponding in the right view **R** can be represented as $\mathbf{R}(x, y')$. The item value in the corresponding position in **Di** is

$$\mathbf{Di}(x, y) = y - y' \tag{7.9}$$

The essential step in disparity estimation is to find the corresponding pixels in a view for the pixels in the other view. For convenience, we will follow the idea to use left view as the basis if not declared otherwise. That is, for any pixel in the left view $\mathbf{L}(x, y)$, its corresponding pixel in the right view is the one matching the following two conditions: The horizontal position of the two pixels is identical, i.e., x is the same; the place that the corresponding pixel in the right view is absolutely to the left to that in the left view, i.e., $y' < y$. Note that although x and y are both integers, y' does not have to be so. The two factors have both been demonstrated in Fig. 7.5. Among the candidate pixels, one that is considered most likely to be the solution is selected as the final result. In this specific paper, three methods are employed to give judgements that which pixel would be the best among the candidates, respectively, based on the simple SAD (Konolige 1997), a segmentation-based method (Scharstein and Szeliski 2002), and SSIM (Wang et al. 2004). The basic idea is to consider the candidate pixel from the right view whose neighbor is mostly similar to that of the corresponding pixel in the left view, and the specific strategies are defining this similarity with different standards.

Since the disparity estimation is an evitable stage in this framework, the depth information is explicitly made use of. Moreover, the depths are connected to specific image contents, rather than separately processed in the similar manner as methods introduced in the previous sub-section. Also, from the prospective of HVS properties, how the depth is employed follows the flow of how people actually perceive stereoscopic image quality: The stereo image pairs are firstly fused into one image, and then the visual cortex judges the quality of the fused image, during which the depth perception is generated in the image fusion process, similar to this objective SIQA flow.

However, a problem occurs in regard to the accuracy of current disparity estimation schemes. Recent years have already witnessed great progress in the pixel matching of binocular views, yet there is no such a method that is capable to achieve perfect accuracy. Especially, when we are dealing with quality problems, the images are usually suffering from distortions, which bring about extra difficulty, especially for no-reference methods. There is actually another concern. In most of the databases established for SIQA, the included images are in the "standard" form, which means the corresponding pixel pairs are ideally on the same horizontal level, yet the assumption often fails in practice, because the cameras are not always set in a uniform manner. That way an extra step is introduced that modifies the views into the uniform form, where again the potential of inaccuracy is caused especially for distorted images.

Disparity estimation is operated for the test and reference images respectively in this work. However, to use only the disparity map estimated by the reference for both the test and reference is also common for full-reference methods (Zhang and Chandler 2015). When it is strongly desired to obtain good disparity estimating results, the latter way is more recommended. But how much this decision affects IQA results is not quite certain. If the disparity map for a distorted stereo image pair is poorly estimated, it suggests a serious distortion and will reflected in the stereoscopic image of the image pair, probably leading to a more significant difference between the disparity map and that of its reference. This situation is not necessarily worse than when the disparity is better computed. For no-reference methods, it is not possible for a distorted stereoscopic image to use the original disparity map, so they cannot expect for ideal disparity maps.

Nevertheless, the disparity estimation is functioning in two ways, at the very least. On one hand, the depth information is explicitly made use of and connected to image contents, overcoming the problem caused by the separative scheme in the last sub-section. On the other hand, it directly paves the path for the next stage, cyclopean image construction. Bad news is that the construction of cyclopean images is, similar to disparity estimation, also an unsolved problem. Biologically, we have no idea how the cyclopean images are really built in our mind, although we are quite certain that it is operated in the very early area in the visual cortex because that is where the visual signals from the two eyes meet each other. In other words, our current knowledge about HVS is still too shallow to guide the generation of a mathematical model that perfectly simulates the process in human brain. Therefore, a popular choice is to replace the complex simulation by simplified mathematical models, among which the most widely adopted one is a linear model proposed in (Levelt 1968). Let **CI** denotes the cyclopean image, and its pixels be all presented as

$$\mathbf{CI}(x,y) = \mathbf{W_L}(x,y)\mathbf{L}(x,y) + \mathbf{W_R}(x,y')\mathbf{R}(x,y') \tag{7.10}$$

where (x, y') is the corresponding point on the right view of the point from the left view, previously found out by disparity estimation. $\mathbf{W_L}$ and $\mathbf{W_R}$ are the weight maps for the left and right views to account for binocular rivalry. Obviously, the equation follows our convention that the process uses left view as the basis. This linear model is adopted by (Chen et al. 2013b) as well as many other SIQA methods. More concretely, the item values in $\mathbf{W_L}$ and $\mathbf{W_R}$ are basically depending on the attention distribution for corresponding regions for the left and right views, which can be computed through an estimation of binocular rivalry. In most cases, we have to ensure that

$$\mathbf{W_L}(x,y) + \mathbf{W_R}(x,y') = 1 \tag{7.11}$$

holds for all image pixels.

From Sect. 7.2.3, we know that the two items on the left end of the above equation both equal to 0.5 when there is a perfect state free of binocular rivalry. When binocular rivalry exists, the balance would be broken. We also know that they can be defined based on the amount of information or energy. Let $\mathbf{E_L}$ and $\mathbf{E_R}$ be the energy maps of the left and right views, we can process

$$\mathbf{W_L}(x,y) = \frac{\mathbf{E_L}(x,y)}{\mathbf{E_L}(x,y) + \mathbf{E_R}(x,y')} \tag{7.12}$$

$$\mathbf{W_R}(x,y') = \frac{\mathbf{E_R}(x,y')}{\mathbf{E_L}(x,y) + \mathbf{E_R}(x,y')} \tag{7.13}$$

to let (7.11) stay ahold.

Although there are other cyclopean image construction models, we do not plan to list all of them because they can hardly be as effective as the linear model, but are usually more computationally complex. The linear model is most widely employed for existing SIQA methods.

Therefore, the last thing we need to do to construct cyclopean images is to find energy maps $\mathbf{E_L}$ and $\mathbf{E_R}$. As is mentioned, they can be determined based on the quantity of visual attention, visual responses or information. Since Gabor filters are regarded as a good simulation of visual responses, they are very common to be employed here, with the form of

$$\mathbf{G}(x, y, \sigma_x, \sigma_y, \zeta_x, \zeta_y, \theta) = \frac{\exp\left(-\frac{1}{2}\left(\left(\frac{R_1}{\sigma_x}\right)^2 + \left(\frac{R_2}{\sigma_y}\right)^2\right) + i\left(x\zeta_x + y\zeta_y\right)\right)}{2\pi\sigma_x\sigma_y} \quad (7.14)$$

where $R_1 = x\cos\theta + y\sin\theta$, $R_1 = y\cos\theta - x\sin\theta$, σ_x and σ_y are the standard deviation of the Gaussian window along axes x and y, ζ_x and ζ_y are spatial frequencies along x and y, and θ denotes the orientation of the filter. A Gabor filter with parameters specifically set is with a certain scale and orientation. Because of the contrast sensitivity property of HVS, the Gabor responses on different scales are not equally important. Usually, four or six orientations are adopted that divides a circle equally, and the most desirable number of scales is not quite clear. Since scale is related to the spatial frequency, which human favors the middle bands, the filters have to be designed to cover the frequencies near the peak of HVS responses. Authors of (Chen et al. 2013b) design the Gabor filter bank following (Su et al. 2011), and the filter bank is with four directions, i.e., horizontal, vertical, and both diagnoses, and only one scale, where the corresponding frequency optimizes HVS responses, although in many other works from IQA constructing Gabor filter with more than one scale is favored (Zhao et al. 2016; Ding et al. 2017).

There exist other approaches appropriate for computing the weights for both views as well. For instances, steerable pyramid and wavelet transform are both capable to extract spatial frequency features. Comparatively, Gabor filters are considered as a better simulation of HVS. Other schemes such as visual saliency can also be employed to compute the weights, because in essence the weights reveal the distribution of visual attention.

It is worth noted that the cyclopean image construction models can be used to fuse not only the two views, but also the filtering responses or extracted features of the views, which is one of the recent trends. For one instance, the method proposed in (Zhou et al. 2017a, b) combines the Laplacian-of-Gaussian responses of the two views by the model similar to (7.3) to simulate the HVS response for the cyclopean image, instead of filtering after constructing the cyclopean image. The two ways are both feasible, and which one is better is really dependent on specific situations.

When the cyclopean images are obtained, the remaining part is merely a 2D IQA problem. If the cyclopean image is ideal, we have reason to believe that proper 2D IQA methods are able to generate desirable predicting results. In (Chen et al. 2013b), four metrics are tried, namely peak signal-to-noise ratio (PSNR), SSIM (Wang et al. 2004), visual information fidelity (VIF), and multiscale SSIM (MS-SSIM) (Wang et al. 2003). The authors have compared the experimental results from the four 2D IQA methods using both the direct averaging scheme, similar to previously introduced in (Campisi et al. 2007), and the cyclopean image construction scheme. As is validated by experiments, the cyclopean image construction generates much better results than the direct averaging, and better 2D methods employed give better results, generally speaking. The effectiveness of the cyclopean image construction-based framework is obvious.

Constructing cyclopean images for SIQA has many benefits. It fundamentally considers the information conveyed by both image content and image depth, as well as the influence caused by binocular rivalry. This is because the flow that constructs cyclopean images and then operates quality assessment upon them is quite alike as the reality in human brain. The framework divides the process of SIQA into serial stages. In each of the stages, mathematical models can be borrowed from related fields in image processing and computer vision.

However, the performance of such methods is easily affected by the accuracy of the disparity estimation or the cyclopean image construction schemes, which are both not stable. On the other hand, these methods are mainly operated in a serial flow, so the computational efficiency could be rather low.

7.4.4 Endowing Weights for Both Views for SIQA

Taking binocular rivalry into account is necessary when designing desirable SIQA methods, and operating 2D methods on cyclopean images is a good way to make it happen, yet generally exhibiting the flaws of instability and inefficiency in the meanwhile. Therefore, researchers come up with the idea of endowing weights for the both views of stereoscopic images to account for binocular rivalry and avoid the cyclopean image construction (Levelt 1965).

Most of the weight-endowing-based methods follow a very simple linear model,

$$Q = W_L Q_L + W_R Q_R \qquad (7.15)$$

where Q_L and Q_R are the quality of the left and right views, and Q is the quality score of the stereoscopic image. Similar to discussions mentioned previously for constructing cyclopean images, many researchers insist that the weights have to be normalized. Usually, it is ensured that

$$W_L + W_R = 1 \qquad (7.16)$$

This model is still in use because it exhibits not only impressive simplicity, but also competitive accuracy for quality prediction. An example of this kind of method is presented previously in (Campisi et al. 2007). In this example, we can find that, for one thing, this model is fairly easy for implementation that it is favored since the very early time, for another, the question how to compute the weights is solved in a very experiential way that is hardly a good solution, offering large room for performance improvement.

Except for the simplicity, the effectiveness of the linear model has also been demonstrated. Therefore, the framework is still in wide use nowadays. Since the quality evaluation is synthesized from results of two aspects, 2D IQA and the weights, the problem can be divided into two and conquer, respectively. Take the full-reference method proposed in (Wang et al. 2015) as an example of modern method of this kind, the framework of which is given in Fig. 7.9. The two parts are explicitly distinguished. Before the weighted averaging, quality scores of left and right views are derived with planar quality assessment metrics, and the weights are computed.

2D IQA methods employed in this work are developed based on SSIM (Wang et al. 2004). Selecting them is not only because of their demonstrated competitive performance, but also that the schemes are pixel-wise, i.e., a quality index is obtained for each image pixel, which would be proved useful later. The quality score of the whole image is the weighted average of these indices. Corresponding to the pixel positions, a quality map is organized with the quality indices as elements, so that the SSIM value of a planar image is the average of the quality map.

The advantage of the fact that the quality indices are derived over space is that it allows us to extend SSIM by defining weights for each quality index, instead of a direct averaging. That is, assuming a weight map is organized at the same time, the elements of which is the weight for the quality indices, the quality score of a planar image can be represented as

$$Q = \frac{\sum_{i=1}^{N} w_i q_i}{\sum_{i=1}^{N} w_i} \tag{7.17}$$

which is actually a pooling stage, w_i and q_i are elements of the weight map and the quality map, respectively, and N is the total number of pixels in the assessed image. Derivates of SSIM, MS-SSIM (Wang et al. 2003), and information content weighting (Wang and Li 2011) are some examples of such extension.

The authors determine weights based on two strategies. The first one is image content, which assumes that the spatial locations containing more information

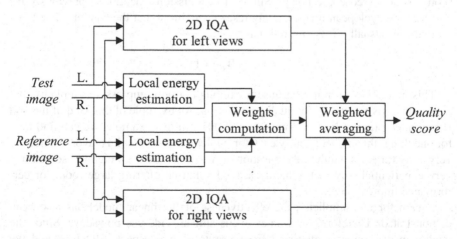

Fig. 7.9 Framework of the method proposed in (Wang et al. 2015)

attract more visual attention, and thus should be given larger values of weights. The information content is evaluated according to (Wang and Shang 2006). A quantitative definition is

$$w_i = \log\left(\left(1 + \frac{\sigma_{xi}^2}{C_1}\right)\left(1 + \frac{\sigma_{yi}^2}{C_1}\right)\right) \tag{7.18}$$

where σ_{xi} and σ_{yi} are the standard deviations of x_i and y_i, the local image patches around the i-th pixel, and C_1 is a hyperparameter called the noisy visual channel power.

The second pooling strategy is based on distortion level, under the assumption that the spatial locations containing more distortions attract more visual attention and acquire larger weights. The distortion level d_i of the i-th pixel is negatively correlated with the local image quality q_i measured previously using SSIM,

$$d_i = 1 - q_i \tag{7.19}$$

And the weight is actually a normalized form of the distortion level,

$$w_i = \frac{d_i}{\sqrt{\sum_j d_j^2 + C_2}} \tag{7.20}$$

where d_j is the distortion level of a pixel in the neighbor of the i-th pixel, and C_2 is another hyperparameter called the stability constant.

Denoting the weight determined for the i-th pixel based on image content and distortion level with w_{i1} and w_{i2}, respectively, the final w_i is computed as

$$w_i = \max\left\{(w_{i1})^2, (w_{i2})^2\right\} \tag{7.21}$$

This newly designed 2D IQA method is referred to as information content and distortion weighted SSIM (IDM-SSIM) by the authors, and is proved significantly advanced comparing to the conventional version of SSIM.

In essence, the weight computation scheme is to convert the 2D IQA method to 3D field, which is accomplished under the guidance of binocular fusion and rivalry theory (Kaufman 1974; Julesz 1971). The specific method employed in this paper is motivated by several previous works (Levelt 1966; Blake 1977; Fahle 1982; Ding and Sperling 2006), which have proved that for a simple visual stimulus, an increasing contrast appearing in a single view will lead to dominance against the other, ideally. In complicated scenarios, the contrast is measured using energy. Thus, binocular rivalry can be quantified by the relative energy of both views.

Firstly, the local energy maps are constructed for both reference and distorted images using a very easy scheme, computing the local variances. The variances are computed with the help of circular-symmetric Gaussian windows of size 11×11 with normalization. Let $E_{r,l}$, $E_{r,r}$, $E_{d,l}$, and $E_{d,r}$ denote the derived local energy maps, where the first subscript "r" and "d" denote the reference and distorted images, and the second subscript "l" and "r" denote the left and right views, respectively. Then, the local energy ratio maps for left and right views, $\mathbf{R}_l$ and $\mathbf{R}_r$, are computed as (with the i-th pixel as an example)

$$\mathbf{R}_l(i) = \frac{\mathbf{E}_{d,l}(i)}{\mathbf{E}_{r,l}(i)} \tag{7.22}$$

$$\mathbf{R}_r(i) = \frac{\mathbf{E}_{d,r}(i)}{\mathbf{E}_{r,r}(i)} \tag{7.23}$$

Obviously, a ratio over 1 indicates the distorted image contains more energy comparing to the reference image, and vice versa.

According to our previous discussion, higher-energy image regions contain more information, attract more visual attention, and are more likely to be in dominance. To pool the maps into one value, the dominance levels of the left and right views are computed by

$$g_l = \frac{\sum_i \mathbf{E}_{d,l}(i) \cdot \mathbf{R}_l(i)}{\sum_i \mathbf{E}_{d,l}(i)} \tag{7.24}$$

$$g_r = \frac{\sum_i \mathbf{E}_{d,r}(i) \cdot \mathbf{R}_r(i)}{\sum_i \mathbf{E}_{d,r}(i)} \tag{7.25}$$

Moreover, it is suggested that the actual dominance level is related to spatial frequency (Fahle 1982; Yang and Blake 1991). To obtain a better simulation according to the contrast sensitivity property of HVS, the images are divided into multiple scales through a set of low-pass filters. The dominance level is a weighted summation of these s scales,

$$g_l = \sum_{i=1}^{s} \alpha_i g_{i,l} \tag{7.26}$$

$$g_r = \sum_{i=1}^{s} \alpha_i g_{i,r} \tag{7.27}$$

where $g_{i,l}$ and $g_{i,r}$ are the dominance levels of scale i for the left and right views, respectively, and α_i is the weight for scale i, which is determined using the CSF model proposed in (Barten 2004),

$$S(u) = \frac{5200 \exp\left(-0.0016u^2\left(1 + \frac{100}{L}\right)^{0.08}\right)}{\sqrt{\left(1 + \frac{144}{X^2} + 0.64u^2\right)\left(\frac{63}{L^{0.83}} + \frac{1}{1 - \exp(-0.02u^2)}\right)}} \tag{7.28}$$

where u is the spatial frequency in cycles/degree, the other two parameters are dependent upon scenarios, L denotes the luminance in cd/m^2, X^2 is the angular object area in square degrees. Using this model, α_i is calculated as

$$\alpha_i = S(f_i) \tag{7.29}$$

where f_i is the central frequency on scale i. The multiscale strategy works because comparing to methods in (Wang and Wang 2014; Wang et al. 2014), which are identical except for the multiscale process, evident performance improvement is witnessed.

Finally, the weights for left and right views are computed with

$$W_L = \frac{g_l^2}{g_l^2 + g_r^2} \tag{7.30}$$

$$W_R = \frac{g_r^2}{g_l^2 + g_r^2} \tag{7.31}$$

And the final quality score is computed using the linear model.

This weighted averaging framework seems rather simple, but exhibits several evident advantages comparing to the cyclopean image construction-based flow. The framework is in a parallel form, so it is generally better in terms of efficiency. Also, the parallel framework contains better extendibility. But, as for the predicting accuracy, it is hard to give a general comparative conclusion for the two frameworks.

Even with the same weight computation and 2D IQA schemes, the performance of the SIQA method can be further improved by at least two ways with slight adjustment on the framework. On one hand, instead of the view-based weight defining approach, i.e., one weight is endowed for each view, block-based and pixel-based strategies might produce better results, although perhaps disparity estimation have to be involved then. On the other hand, machine learning techniques can be employed to replace the linear regression function, to allow the SIQA algorithms to "learn" how to map the 2D quality of the views to the final quality score. This is another major difference comparing to the cyclopean image construction-based methods, because although we can find a good explanation for the linear model according to binocular rivalry effect, the model construction is actually more like a mathematical problem, which is not as necessary for requiring biological basis as constructing the cyclopean images.

Similar to our explanations in Sect. 7.4.3 for cyclopean image construction-based methods, the weight-endowing-based methods contain many derived variations as well that do not strictly follow the framework as Fig. 7.9. For one instance, the weight computation scheme in (Wang et al. 2015) makes use of both the information from the tested and reference stereoscopic image, but it is also unnecessary. For no-reference methods, it is only possible to derive the weights from the tested image pairs, and this is feasible because it is the relative energy that causes the different weight distribution, not the absolute energy; and this is certainly as practical as for full-reference methods. To summarize, weight-endowing is merely a thought for SIQA, and how it should be specifically applied in practice varies according the ideas of algorithm designers as long as it is theoretically correct and practically effective.

7.5 Combinational Solutions for Objective SIQA

7.5.1 Theoretical Basis for Combinational Thoughts

The above-discussed frameworks are all quite straightforward, but current SIQA methods tend to be more complicated, especially by combining different frameworks or feature extraction schemes together, for the pursuit of obtaining better predicting accuracy. Listing all the combinations that existing works have tried is impossible, yet some basic principles and favorable trends can be observed.

The assumption that combinations improve performance is based on the idea that different approaches might extract complementary features from the stereoscopic images. According to our previous analysis, feature extraction schemes are classified as three types: from the both views, from the disparity maps, and from the cyclopean images. Of course, features from the three different sources carry different information and might make good complements for each other.

Extracting features from the left and right views can be done easily by applying many thoroughly studied feature extraction tools that are proved efficient for 2D IQA. This way we are sure that the captured information is what the left and right eyes are interested in, but with the influence of binocular rivalry, we are not certain about whether the captured information is really attractive to the visual cortex.

However, we are really not sure how strong features from the disparity maps affect the perception upon stereoscopic image quality. What we do know is that the distortion of stereoscopic images reflects in the damage of their disparity maps (Benoit et al. 2008b) and the degree of visual discomfort is related to the disparity distribution (Lambooij et al. 2009). Generally speaking, features from the disparity maps are worth studying. An advantage is that if there is useful information in the disparity maps, it seems not to overlap too much with the information conveyed by either the both views or the cyclopean images.

Theoretically, information contained in cyclopean images correlates best with quality perception among the three, because how HVS operates subjective SIQA is very much similar to conducting 2D IQA on the cyclopean images constructed in our minds. Binocular effects, including at least binocular rivalry, are inherently taken into consideration. The problem is that the cyclopean images cannot be constructed ideally.

As a result, each of the feature type possesses merits and flaws, so the combinational thought might be promising. If we employ machine learning to train regression function that synthesizes the quality indices computed using multiple features, the feature extraction processes can be organized all in a parallel framework, and the computational time would only depend on the slowest one. With the popularization of machine learning and distributed computing tools, breakthroughs in objective SIQA are certainly most likely to occur in the form of combinational methods. In the following sub-sections, we will introduce four state-of-the-art examples that operate SIQA with complex and combinational approaches.

7.5.2 3D-MAD Based on Binocular Lightness and Contrast Analysis

The first example is called 3D-MAD (most apparent distortion) (Zhang and Chandler 2015), of which the framework is given in Fig. 7.10. As the name implies, it is an SIQA method built based, to some extent, upon the 2D method MAD (Larson and Damon 2010), determining that the method is a full-reference one. As the authors define, the computed objective quality score for stereoscopic images is the product of two quality scores derived by different means, called "2D-MAD" score and "Cyc-MAD" score. The former is derived with the framework similar to the case in (Wang et al. 2015) discussed in Sect. 7.4.4, of course with evident difference in the feature extraction and weight computation schemes; the latter is a cyclopean image construction-based SIQA method, similar to what we introduced in Sect. 7.4.3. More detailed implementation will be introduced in the following.

In the 2D-MAD quality score computation stage, the conventional MAD algorithm is operated on both views separately, only with the trained parameters adjusted. The MAD is a product of two indices, the perceived distortion due to visual detection (d) and the perceived distortion due to visual appearance (a), computed in the form of

$$MAD = d^{\alpha} \cdot a^{1-\alpha} \qquad (7.32)$$

where α is a parameter between 0 and 1 to determine the relative importance between d and a,

$$\alpha = \frac{1}{1 + \beta_1 \cdot d^{\beta_2}} \qquad (7.33)$$

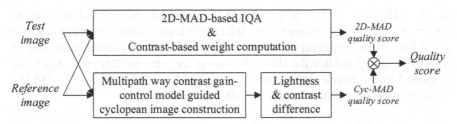

Fig. 7.10 Framework of the full-reference method in (Zhang and Chandler 2015)

The specific values of β_1 and β_2 are slightly different from that in the traditional 2D MAD because the selected training dataset is LIVE 2D database (Sheikh et al. 2006) instead of A57 database (Chandler and Hemami 2007) whole scale is too small. How to define d and a is quite complex and is identical to the method described in 2D MAD (Larson and Damon 2010), thus will not be discussed in detail here. Note that values of β_1 and β_2 are both scale-dependent, and five scales are constructed.

Then, weights are determined for both views so that the 2D MAD can serve for 3D purposes. The authors explain binocular rivalry effect by the concept "masking" and quantify that by calculating local contrast. So, a block-based contrast map is organized for each view. Concretely, the image is firstly divided into blocks of size 16×16 with 75% overlapping between neighboring blocks, and then the root mean squared contrast of each block is measured,

$$C(b) = \frac{\sigma(b)}{\mu(b)} \tag{7.34}$$

where $\mu(b)$ is the average luminance of block b, and $\sigma(b)$ is the minimum standard deviation of the four 8×8 sub-blocks in b.

Four contrast maps are constructed for the left and right views (denoted as subscripts "L" and "R") of the reference and distorted images (denoted as superscripts "r" and "d"), i.e., $\mathbf{C}_L^d, \mathbf{C}_L^r, \mathbf{C}_R^d$ and $\mathbf{C}_R^r$. The weights for left and right views, W_L and W_R, are determined by

$$W_L = \left(\frac{\sum \mathbf{C}_L^d}{\sum \mathbf{C}_L^r}\right)^{\gamma} \tag{7.35}$$

$$W_R = \left(\frac{\sum \mathbf{C}_R^d}{\sum \mathbf{C}_R^r}\right)^{\gamma} \tag{7.36}$$

where Σ denotes the summation of all elements in a map, and γ is set to 2 so that higher-contrast regions are emphasized.

The final 2D-MAD score S_{2D} is computed by

$$S_{2D} = \frac{W_L \cdot \exp\left(\frac{MAD_L}{100}\right) + W_R \cdot \exp\left(\frac{MAD_R}{100}\right)}{W_L + W_R} \tag{7.37}$$

where MAD_L and MAD_R are the 2D MAD scores of the left and right views. Obviously, the above equation is very alike to the linear synthesis model introduced in the last sub-section, except for that it integrates the weight normalization into the final pooling step and that the 2D quality scores are exponentially processed before employed for generating the 3D scores. Thus, the flow of computing the 2D-MAD scores is basically the same as the weight-based strategy.

The path parallel to the 2D-MAD computation is finding the *Cyc-MAD* scores, which is similar to the cyclopean image construction-based SIQA flow. The difference between the scheme adopted here and the conventional cyclopean image-based one is that the cyclopean feature maps are built by fusing the feature maps of both views, rather than fusing views to cyclopean images.

There are two kinds of features employed here, the lightness distance and the contrast. Specifically, global and local lightness distance maps are both constructed for comprehensive description, and the contrast is measured in a pixel-based manner.

For the *i*-th pixel, the first feature, global lightness distance, is determined as

$$\mathbf{F}_1(i) = \left| \mathbf{L}(i) - \frac{1}{N} \sum_{j=1}^{N} \mathbf{L}(j) \right| \tag{7.38}$$

where N is the number of pixels in the image, and $\mathbf{L}$ is the lightness component in the Commission Internationale de l'Eclairage (CIE) 1976 color space. The local lightness distance, as well as the second features, is defined as the difference between $\mathbf{L}(i)$ and the average in a neighbor $\mathbf{B}$ with size of 9×9,

$$\mathbf{F}_2(i) = \left| \mathbf{L}(i) - \frac{1}{81} \sum_{j=1}^{81} \mathbf{L_B}(j) \right| \tag{7.39}$$

where the subscript $\mathbf{B}$ denotes that the *j*-th pixel belongs to neighbor $\mathbf{B}$.

The third feature is the pixel-based contrast defined as

$$\mathbf{F}_3(i) = \frac{\mathbf{L}(i)}{\frac{1}{81} \sum_{j=1}^{81} \mathbf{L_B}(j) + K} \tag{7.40}$$

where $K = 0.001$ is a small constant to prevent the denominator from being 0. The three feature maps are all built from raw binocular views.

Certainly, constructing disparity maps is another inevitable step, which is accomplished by the segment-based stereo-matching approach presented in (Klaus et al. 2006). With the monocular feature maps and the disparity maps, the cyclopean feature maps can be developed. Making use of feature maps is accomplished by a multipathway contrast gain-control model (MCM) (Huang et al. 2010, 2011), which assumes that the contrast gain control is depending not only on the visual signal of the other eye, but also the incoming gain-control signal from the other eye.

Concretely, three gain-control mechanisms are employed. Specifically, there exist attenuation of signal contrast in the nondominant eye, a stronger direct inhibition of the dominant eye, and a stronger indirect inhibition from the dominant eye comparing to the gain-control signal of the nondominant one. The original scene contrast is previously derived as $\mathbf{C}_L$ and $\mathbf{C}_R$, and after the simulation of the gain-control mechanisms, the signal strengths by the eyes are

$$C_L' = \frac{C_L}{1 + \frac{\varepsilon_R}{1 + \beta \varepsilon_L}} \tag{7.41}$$

$$C_R' = \frac{C_R}{1 + \frac{\alpha \varepsilon_L}{1 + \varepsilon_R}} \tag{7.42}$$

where

$$\varepsilon_L = \rho \mathbf{C}_L^{\gamma_1} \tag{7.43}$$

$$\varepsilon_R = \rho (\eta \mathbf{C}_R)^{\gamma_1} \tag{7.44}$$

The above equations are under the assumption that the left-eye dominants, parameter ρ denotes the contrast gain-control efficiency, η models the contrast attenuation in the right eye, α and β simulate the stronger inhibition to the right eye from the left one, and γ_1 offers nonlinearity of the gain control. Then, the cyclopean feature map is given by

$$\mathbf{C} = \left((\mathbf{C}_L')^{\gamma_2} + (\mathbf{C}_R')^{\gamma_2} \right)^{\frac{1}{\gamma_2}} \tag{7.45}$$

where γ_2 offers nonlinearity of the binocular contrast combination.

Taking into consideration both the disparity and the MCM cyclopean construction technique, we can obtain a general form of the cyclopean ($\mathbf{C}$) of all the three feature maps ($\mathbf{F}_1$, $\mathbf{F}_2$, $\mathbf{F}_3$), i.e., for the i-th pixel whose coordinates are (x, y), denote its disparity as d_i, $\mathbf{C}$ can be represented as

$$\mathbf{C}_{F_j}(x, y) =$$

$$\frac{\left(\left(\eta_L \mathbf{F}_{i,L}(x - d_i, y) \frac{1}{1 + \frac{\varepsilon_R(x - d_i, y)}{1 + \beta \varepsilon_L(x - d_i, y)}} \right)^{\gamma_2} + \left(\eta_R \mathbf{F}_{i,R}(x, y) \frac{1}{1 + \frac{\alpha \varepsilon_L(x, y)}{1 + \varepsilon_R(x, y)}} \right)^{\gamma_2} \right)^{\frac{1}{\gamma_2}}}{\left(\left(\eta_L \frac{1}{1 + \frac{\varepsilon_R(x - d_i, y)}{1 + \beta \varepsilon_L(x - d_i, y)}} \right)^{\gamma_2} + \left(\eta_R \frac{1}{1 + \frac{\alpha \varepsilon_L(x, y)}{1 + \varepsilon_R(x, y)}} \right)^{\gamma_2} \right)^{\frac{1}{\gamma_2}}} \tag{7.46}$$

which holds for $j = 1, 2, 3$.

The quality is represented as the local statistical differences between the cyclopean feature maps of the reference and distorted stereoscopic images, including the standard deviation, skewness, and kurtosis computed for a 16-by-16 block b with 75% area overlapping, which is done by decomposing the stereoscopic images into 5 scales and 4 orientations using a bank of log Gabor filters and computing the three difference maps on each scale, specifically. On each scale, the difference is represented as

$$\eta(b) = \sum_{o=1}^{4} \left| \varphi_o^{\text{ref}}(b) - \varphi_o^{\text{dis}}(b) \right| \tag{7.47}$$

where φ denotes either of the three statistical features, $o = 1\text{–}4$ is the orientation, and the subscript "ref" or "dis" specifies whether the features are from the reference image or distorted image. The statistical difference map is pooled into a single scalar φ_s,

$$\varphi_s = \left(\frac{\sum_b \eta(b)^2}{B_\varphi} \right)^{\frac{1}{2}} \tag{7.48}$$

where $s = 1\text{–}5$ denotes the scale, and B_φ is the total number of blocks within one statistical difference map at scale s. As a result, for all the five scales, three quality indices are generated. That is, for either of $\mathbf{F}_1$, $\mathbf{F}_2$, and $\mathbf{F}_3$, 15 indices are developed, respectively.

For the local and global lightness distance cyclopean feature maps ($\mathbf{F}_1$ and $\mathbf{F}_2$), the 15 indices are mapped into one quality score (S_1 and S_2) by support vector machine training on LIVE 2D database (Sheikh et al. 2006). For the pixel-based contrast feature maps, a strategy similar to what is presented in (Larson et al. 2010) is adopted. Only the medium scale ($s = 3$) is used, and absolute difference of the three statistics (standard deviation σ, skewness ς, and kurtosis κ) is weighted summarized,

$$\eta(b) = \sum_{o=1}^{4} \left(\left| \sigma_{3,o}^{\text{ref}}(b) - \sigma_{3,o}^{\text{dis}}(b) \right| + 2 \left| \varsigma_{3,o}^{\text{ref}}(b) - \varsigma_{3,o}^{\text{dis}}(b) \right| + \left| \kappa_{3,o}^{\text{ref}}(b) - \kappa_{3,o}^{\text{dis}}(b) \right| \right) \tag{7.49}$$

and

$$S_3 = \left(\frac{\sum_b \eta(b)^2}{B} \right)^{\frac{1}{2}} \tag{7.50}$$

The quality score S_{cyc} computed using the cyclopean feature map construction way is

$$S_{\text{cyc}} = \frac{S_1 + S_2}{10} + S_3 \tag{7.51}$$

Finally, the quality score S of a stereoscopic image is given as

$$S = S_{2D} \cdot S_{cyc} \tag{7.52}$$

This method is a typical example among the combinational methods we refer to. It is worthwhile that we analyze it in detail, to discover why such a framework is capable to include multiple significant innovations and to result in satisfactory outcome.

Firstly, let's take a look at how 2D IQA methods are introduced for the 3D problem. In the upper path of the flowchart where 2D-MAD scores are computed, MAD is applied to compute the respective quality scores for both views, with different trained parameters from the conventional 2D MAD. In the lower path, the statistical differences between the lightness distance and pixel-based contrast are employed upon the constructed cyclopean images. Lightness, contrast, and MAD have all been proved effective for 2D IQA and are reasonably useful for stereo image pairs as well. On the other hand, the processing stages that these 2D methods are made use of are exactly what we have concluded in previous sections: for the respective views and for the cyclopean images. However, the authors have adopted different approaches for these two purposes, which probably suggest that the proper 2D IQA methods for the different evaluated objects can be different and thus should be wisely selected.

Secondly, how the 2D methods are transferred for 3D purposes is varied. A general idea running through the paper is that the eye receiving visual signals with higher contrast dominates, i.e., binocular rivalry is explained based on contrast computation. The weights for left and right views are derived by block-based contrast estimation, and the cyclopean image is constructed based on gain-control mechanisms developed according to signal contrast through various ways. Using contrast to account for binocular rivalry is very alike to the information or energy concept, the latter two are referred to more often, but the essence is basically the same. We have observed many SIQA methods that adopt energy computation schemes such as Gabor filters, but technically, Gabor responses can also be interpreted as local gradient or contrast, and cannot be regarded as a perfectly accurate simulation for the eye dominance or binocular rivalry. Therefore, it should be encouraging to discover more feasible ways to give us more choices to model binocular effects and to find out which way is better.

Last but not least, multiple synthesizing schemes that combine the quality scores derived from various ways are adopted in the work. Naturally, there has to be a way to fuse *2D-MAD* and *Cyc-MAD* scores, and the authors come up with the idea to directly multiply them. Because they are both negatively related to image quality and use value 0 to denote undistorted images, the multiplying trick is feasible. Certainly, many other methods can be adopted, but the convenience and simplicity of the multiplication are obvious. Tracing back to the computation of the two scores, *2D-MAD* is the weighted average of MAD scores of left and right views, and *Cyc-MAD* is the weighted average of three indices, statistical reference between test and reference images representing global lightness distance, local lightness distance, and pixel-based contrast. Focusing on the weights, we can notice that the weights for *2D-MAD* are computed with a contrast-based scheme, and that for *Cyc-MAD* is empirically fixed, but neither of them is trained by machine learning. Although machine learning is also feasible to find the weights, the complexity is again avoided. The only place that more complex synthesizing scheme is introduced is the computation of the global and local lightness distance statistical difference between the tested images and their reference, where support vector regression (SVR) is employed.

Based upon the above analysis, why the combinational SIQA methods can be rich in content and satisfactory in performance is becoming clear. Existing simple frameworks, such as the cyclopean image construction-based and the weight-based ones, can be composed together with various modifications and adjustments according to the algorithm designers' demands, and successful 2D IQA methods are always welcomed to be embedded in whatever frameworks. The above example also reveals a novel problem that the combinational SIQA methods are likely to be much more complex than 2D IQA method, so the trade-off between effectiveness and efficiency should also be considered, both in feature extraction and score pooling and synthesis.

7.5.3   Deep Quality Estimator Based on Monocular and Binocular Interactions

The above-mentioned factors in regard to the design of objective SIQA methods can also be found in other methods; we will introduce (Shao et al. 2016a, b) called 3D-DQE (deep quality evaluator) as the second example of modern combinational methods in the following. The reason we would like to introduce it is manifold. To name a few, unlike all the previous discussed methods in this chapter, it is a no-reference one, which might give us some new insights for SIQA; it adopts deep learning networks, which is uncommon in IQA research. Actually, as the authors themselves claim, only a few previous works have employed deep learning for IQA (Tang et al. 2014; Ghadiyaram and Bovik 2014; Hou et al. 2015; Kang et al. 2014; Gu et al. 2014).

The framework of the method is given in Fig. 7.11. Clearly, two deep belief nets (DBNs) are trained for the monocular views and the cyclopean images, and are called DBN-A and DBN-B, respectively. For both the monocular views and the cyclopean images, the features are represented as the combination of gradient magnitude, difference of Gaussian, and local binary pattern; and the cyclopean images are constructed with GM and DoG to account for binocular rivalry, denoted as C1 and C2, respectively. The outputs of the DBNs include not only the quality scores q, but also "deep feature" g, which is used to compute weights when the two quality scores computed using either DBN-A or DBN-B are combined (i.e., Combination 1 and Combination 2). The final quality scores are computed through a linear combination of results from Combination 1–3. More detailed introduced is given in the following paragraphs.

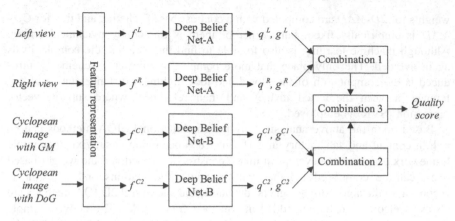

Fig. 7.11 Framework of the no-reference SIQA method proposed in (Shao et al. 2016a, b)

The theory of 3D-DQE is based on the assumption that the quality of a stereoscopic image is a combination of the quality evaluation that is predicted by the 2D monocular deep quality evaluator (2D-DQE) and the cyclopean-based binocular deep quality evaluator (Cyc-DQE). The idea to construct deep networks is based on their successful records in 2D IQA and is aiming to reveal the possible interactions between monocular and binocular visions. Through this way, the problem of studying cyclopean images degenerates to that of studying monocular views or cyclopean images, which are both 2D images. Thus, deep networks applicable for 2D images have the potential to be employed here. The 2D-DNN (deep neural network) is a multilayer DBN, as mentioned earlier, which learns a regression function after training in a dataset, similar to what is implemented in (Ghadiyaram and Bovik 2014) for 2D IQA.

Inspired by (Xue et al. 2014; Shao et al. 2015b; Wu et al. 2015), three features are adopted to describe image quality, which are the difference of Gaussian (DoG), gradient magnitude (GM), and local binary pattern (LBP), extracting image features from different aspects.

The DoG feature of an image $\mathbf{I}$ is generated by computing the difference between its two filtering results using two Gaussian filters with different parameters. Supposing the standard deviation of the Gaussian kernel is σ, the DoG feature map is computed by

$$\mathbf{D} = \mathbf{I} * \mathbf{h} \tag{7.53}$$

$$
\begin{aligned}
\mathbf{h}(x, y | \sigma) &= G(\sigma) - G()k\sigma \\
&= \frac{1}{2\pi\sigma^2}\exp\left(-\frac{x^2+y^2}{2\sigma^2}\right) - \frac{1}{2\pi k^2\sigma^2}\exp\left(-\frac{x^2+y^2}{2k^2\sigma^2}\right)
\end{aligned}
\tag{7.54}
$$

where * is the convolution symbol, G is the Gaussian function, and k is a constant that assists σ to compose the DoG pyramid (Lin and Wu 2014). The DoG features can be represented as the estimated parameters of generalized Gaussian distribution (GGD) on each scale, i.e., the magnitude, variance, and entropy (He et al. 2012; Mittal et al. 2012a). With a five-scale pyramid constructed, 35 DoG features are obtained for an image.

The GM feature is computed as

$$G = \sqrt{(\mathbf{I} * \mathbf{h}_x)^2 + (\mathbf{I} * \mathbf{h}_y)^2} \tag{7.55}$$

$$
\begin{aligned}
\mathbf{h}_x(x,y) &= \frac{x}{2\pi\sigma^4}\exp\left(-\frac{x^2+y^2}{2\sigma^2}\right) \\
\mathbf{h}_y(x,y) &= \frac{y}{2\pi\sigma^4}\exp\left(-\frac{x^2+y^2}{2\sigma^2}\right)
\end{aligned}
\tag{7.56}
$$

With 2 GGD fitting parameters and 16 asymmetric GGD (AGGD) fitting parameters on each scale derived using the method proposed in (Saad et al. 2012), the GM features are of size 42 in total.

The LBP feature is set to radius 1 and neighbor 8, and the LBP histogram of each patch is developed to represent the texture invariance (Ojala et al. 2002). Thirty-six features are generated. Therefore, overall, we have a feature vector of length 113. Mapping the feature vector to the final quality score can be directly done by regression model training such as linear regression, support vector machine, and shallow neural networks. However, because 113 features are generated for either of the monocular view and the cyclopean image, the dimension of raw features would be too high for training shallow models. In addition, the equipment of DNN allows for developing more generalized feature representation than the raw features. Therefore, in the following represents the essential part of this method, the network learning, which maps the raw features of three kinds, $\mathbf{f}_{DoG}$, $\mathbf{f}_{GM}$, and $\mathbf{f}_{LBP}$, to a new feature vector represented by $\mathbf{g}$,

$$\mathbf{g} = DBN(\mathbf{f}_{DoG}, \mathbf{f}_{GM}, \mathbf{f}_{LBP}) \tag{7.57}$$

Two 2D DBNs are used to represent the features in an efficient and effective way (Bengio 2009; Mohamed et al. 2011), which both contain three hidden layers (Hinton et al. 2006). Let the original features be denoted as vector $\mathbf{f}$, and the three hidden layers be denoted as $\mathbf{h}_1$, $\mathbf{h}_2$, and $\mathbf{h}_3$; the joint distribution of them can be represented as

$$P(\mathbf{f}, \mathbf{h}_1, \mathbf{h}_2, \mathbf{h}_3) = P(\mathbf{f}|\mathbf{h}_1)P(\mathbf{h}_1|\mathbf{h}_2)P(\mathbf{h}_2|\mathbf{h}_3)P(\mathbf{h}_2, \mathbf{h}_3) \tag{7.58}$$

The DBN is trained by two serial steps, the unsupervised pretraining phase and the supervised fine-tuning phase. Every two adjacent layers of a DBN are initialized as a restricted Boltzmann machine (RBN) (Yamashita et al. 2014). Specifically, because the prior features are continuous data, the first layer is modeled by a Gaussian–Bernoulli RBM (GB-RBM), while the rest layers are by Bernoulli–Bernoulli RBMs (BB-RBMs), because the neurons are all in binary form. The GB-RBM is consisted of two layers, $\mathbf{f}$ and $\mathbf{h}_1$, so the conditional probabilities for neurons in the two layers are

$$P(\mathbf{f}|\mathbf{h}_1) = \prod_i N\left(b_i + \sigma_j^2 \sum_j h_j w_{ij}\right) \tag{7.59}$$

$$P(\mathbf{h}_1|\mathbf{f}) = \prod_j f\left(c_j + \sum_i \frac{v_i w_{ij}}{\sigma_i^2}\right) \tag{7.60}$$

where N represents the Gaussian density function, f represents the sigmoid activation function, b_i and c_i are biases corresponding to visible and hidden neurons, w_{ij} is the weight between the i-th neuron in $\mathbf{f}$ and the j-th neuron in $\mathbf{h}_1$, and σ is the standard variance of the visible neurons. The specific parameters are trained through an iterative process.

For the BB-RBMs, take the first and second hidden layers as example, and the estimated probabilities are

$$P(\mathbf{h}_1|\mathbf{h}_2) = \prod_i f\left(b_i + \sum_j h_j w_{ij}\right) \tag{7.61}$$

$$P(\mathbf{h}_1|\mathbf{h}_2) = \prod_j f\left(c_j + \sum_i v_i w_{ij}\right) \tag{7.62}$$

The numbers of neurons in $\mathbf{h}_1$, $\mathbf{h}_2$, and $\mathbf{h}_3$ are 200, 150, and 100, respectively. Therefore, with the help of the pretraining phase, the 113 prior features are now represented by 100 values.

The second stage, supervised fine-tuning phase, aims to minimize the gap between the prediction and ground-truth data. Let $\mathbf{g}$ be the output of the third hidden layer $\mathbf{h}_3$, and y be the ground-truth data (here y is the subjective quality score). For a training dataset containing m samples, the training target is to continuously minimize the cost function

$$J = \arg\min \sum_{i=1}^{m} (y_i - \varphi(\mathbf{g}_i))^2 \tag{7.63}$$

where the subscript i denotes that specific $\mathbf{g}$ or y is that of the i-th sample. The function φ that maps $\mathbf{g}$ to y has the form of

$$\varphi(\mathbf{g}) = \sum_{i=1}^{m} \mathbf{w}_i \cdot K(\mathbf{g}, \mathbf{g}_i) + b \qquad (7.64)$$

where $\mathbf{w}_i$ and b are the weight vector and bias constant for the linear regression function that maps the results of $K(\mathbf{g}, \mathbf{g}_i)$ to $\varphi(\mathbf{g})$. The kernel function K has the specific form that can be represented as

$$K(\mathbf{g}_i, \mathbf{g}_j) = \exp\left(-\frac{\|\mathbf{g}_i - \mathbf{g}_j\|^2}{\gamma^2} \right) \qquad (7.65)$$

which is a Gaussian kernel whose width is controlled by γ. The weights and biases are also trained iteratively. To be more specific, a back-propagation process is implemented here to train the parameters.

The two 2D DNNs (DBN-A and DBN-B, as explained previously) are trained in the same manner, and both with the subjective image quality as output label. The difference is that the DBN-A is trained using the monocular view from the dominant eyes, and the inputs of DBN-B are the cyclopean images. The reason that only the dominant views are employed for training is that the training target is actually a combination of both eyes, thus the dominant eye is comparatively more feasible for the sake of visual inhibition. The cyclopean images are constructed using the DoG responses as inputs.

With the trained DBN-A, in the testing stage, monocular views from both eyes are made use of. Denoting $\mathbf{f}_L$ and $\mathbf{f}_R$ as the raw feature vectors of left and right views, now we know that DBN-A can develop feature representation $\mathbf{g}$ and predicting quality score q for both views,

$$\{\mathbf{g}_L, q_L\} = \text{DBN} - \text{A}(\mathbf{f}_L) \qquad (7.66)$$

$$\{\mathbf{g}_R, q_R\} = \text{DBN} - \text{A}(\mathbf{f}_R) \qquad (7.67)$$

The stereoscopic quality score predicted based on monocular views, q_M, is the weighted sum of the quality of the left and right views with $\mathbf{g}_L$ and $\mathbf{g}_R$ acting as the weights,

$$q_M = w(\mathbf{g}_L) \cdot q_L + w(\mathbf{g}_R) \cdot q_R \qquad (7.68)$$

where w is the function that defines weights according to the distribution of feature representation $\mathbf{g}$.

How the function w should be designed is also an important problem. The prior distributions $P(\mathbf{g}_L)$ and $P(\mathbf{g}_R)$ are modeled as

$$P(\mathbf{g}_L) \propto \|\mathbf{g}_L\|^2 \qquad (7.69)$$

$$P(\mathbf{g}_R) \propto \|\mathbf{g}_R\|^2 \qquad (7.70)$$

And the joint distribution $P(\mathbf{g}_L, \mathbf{g}_R)$ follows

$$P(\mathbf{g}_L, \mathbf{g}_R) \propto \langle \mathbf{g}_L, \mathbf{g}_R \rangle \tag{7.71}$$

where the symbol <**a**, **b**> denotes the inner product of the distributions **a** and **b**.

Therefore, the overall weight function is determined as

$$w(\mathbf{g}_R) = \frac{\|\mathbf{g}_R\|^2 + C}{\langle \mathbf{g}_L, \mathbf{g}_R \rangle + C} \tag{7.72}$$

$$w(\mathbf{g}_L) = \frac{\|\mathbf{g}_L\|^2 + C}{\langle \mathbf{g}_L, \mathbf{g}_R \rangle + C} \tag{7.73}$$

where C is a small constant to prevent the denominator from being 0.

The equation that uses quality of both the monocular views and the weights to compute q_M seems very similar. It can be regarded as a simulation of binocular rivalry because the two basic principles are still loyally followed: The weights for both views are equal to 0.5 at the same time when $\mathbf{g}_L$ and $\mathbf{g}_R$ are identical, if not, the monocular view with higher gain (higher $\|\mathbf{g}\|$) dominants (Levelt 1965). But the difference of this equation here comparing to common models lies in the fact that the sum of the weights does not always equal to 1. A little derivation here will help understand why this still works.

Let's take a look at the inverse of the sum of the two weights, and we multiply it by 2 for convenience in explanation,

$$
\begin{aligned}
\frac{2}{w(\mathbf{g}_L) + w(\mathbf{g}_R)} &= \frac{2\langle \mathbf{g}_L, \mathbf{g}_R \rangle + C}{\|\mathbf{g}_L\|^2 + \|\mathbf{g}_R\|^2 + C} \\
&\approx \frac{2\|\mathbf{g}_L\| \cdot \|\mathbf{g}_R\| + C'}{\|\mathbf{g}_L\|^2 + \|\mathbf{g}_R\|^2 + C'} \cdot \cos\theta \\
&= S \cdot \cos\theta
\end{aligned} \tag{7.74}
$$

where C' is another small constant, θ is the angle between feature vectors $\mathbf{g}_L$ and $\mathbf{g}_R$, and S represents the contrast comparison between the two vectors,

$$S = \frac{2\|\mathbf{g}_L\| \cdot \|\mathbf{g}_R\| + C'}{\|\mathbf{g}_L\|^2 + \|\mathbf{g}_R\|^2 + C'} \tag{7.75}$$

Evidently, S quantifies the similarity between $\|\mathbf{g}_L\|$ and $\|\mathbf{g}_R\|$, ranging from 0 to 1. If $\mathbf{g}_L$ and $\mathbf{g}_R$ are the same, S would be equal to 1; the more $\mathbf{g}_L$ and $\mathbf{g}_R$ are different to each other, S becomes smaller. So, the weighted averaging scheme employed in this method is feasible in theory.

For the cyclopean image-based quality estimation, GM and DoG features are adopted as the measurement of energy during the construction of cyclopean images, instead of the most commonly employed Gabor features (Bensalma and Larabi 2013). This strategy is very clever because the GM and DoG features have both been calculated previously; this data re-utilization saves both computational time and hardware resources.

Upon obtaining the trained DBN-B, the feature representation and quality scores for the cyclopean images can be derived,

$$\{\mathbf{g}_{C_1}, q_{C_1}\} = \mathrm{DBN} - \mathrm{B}(\mathbf{f}_{C_1}) \tag{7.76}$$

$$\{\mathbf{g}_{C_2}, q_{C_2}\} = \mathrm{DBN} - \mathrm{B}(\mathbf{f}_{C_2}) \tag{7.77}$$

where subscripts C_1 or C_2 denote that the cyclopean image is constructed with energy maps estimated using GM or DoG features.

The quality score predicted from the binocular vision, denoted as q_B, is defined as

$$q_B = \frac{\|\mathbf{g}_{C_1}\|^2 \cdot q_{C_1} + \|\mathbf{g}_{C_2}\|^2 \cdot q_{C_2}}{\|\mathbf{g}_{C_1}\|^2 + \|\mathbf{g}_{C_2}\|^2} \tag{7.78}$$

This is also a simulation for gain control.

With the obtained quality scores from monocular and binocular visions, q_M and q_B, the remained problem is to map them to the final quality score q. A simple linear model is adopted, in the form of

$$q = q_M - \beta \cdot q_B \tag{7.79}$$

where the parameter β is to be determined. The equation is different from simple linear equations, for its containing physical explanations that the purpose of the positive value β is to eliminate the possible overlapping between q_M and q_B. Training is based on Ningbo University 3D IQA Database (Wang et al. 2009; Zhou et al. 2011) by grid searching for the optimization of Spearman's rank-order correlation coefficient between objective and subjective quality evaluation.

This framework is interesting because the machine learning procedure is set at an early stage, instead of employed for final quality score pooling as in most machine learning-based IQA works. On the contrary, in the pooling stage, which can be roughly considered as the union of Combination processes 1, 2, and 3, there is no involvement of complex learning-based training models (here we do not regard the parameter defining in Combination 3 is a machine learning-based training because the parameter is pretrained and fixed, using the grid search scheme). Still, this approach is inspired by the idea to build a connection between the features and the quality scores.

Despite of the complexity, we can easily observe that the framework can be regarded as a composition of two parallel parts: a weight computation-based SIQA method (the DBN-A-based paths) and a cyclopean image-based method (the DBN-B-based paths). Therefore, it can be safely stated that, even with our currently shallow understanding upon human binocular vision, there are a considerable number of untried approaches for innovations and improvements in SIQA study.

7.5.4 Saliency-Guided Binocular Feature Consolation

The third example we would like to discuss is another no-reference method presented in (Xu et al. 2017), the flow of which is given in Fig. 7.12. In their paper, the cyclopean image construction-based methods are criticized for their complexity and inefficiency (Shao et al. 2016a, b; Zhang et al. 2016). Therefore, the main framework is basically based on the weight computation. The uniqueness of the method from conventional weight computation-based methods lies in that the features from disparity maps are extracted and considered as quality indices in the meantime. Because the disparity map derivation and the corresponding feature extraction and quantification are in a parallel path to the weight computation flow, rather than the cyclopean image-based methods that the disparity maps are utilized in further process, the total computational time is not increased. Moreover, many novel ideas are applied to make their strategy more convincing and perform better, such as the complementary feature extraction using global and local features, the employment of multiscale saliency maps, and the schemes to capture features from disparity maps.

To the pursuit of a comprehensive quality description, the authors try to capture the quality-aware features from different scales and aspects. It is assumed that the global and local features make good compensation for each other. Therefore, the proposed solution is to capture multiscale global and local features.

Global features are extracted in the spatial domain. In the first, the luminance is converted to log-contrast representation through a local nonlinear transform, which eliminates the local mean displacements from zero log-contrast. Then, the mean

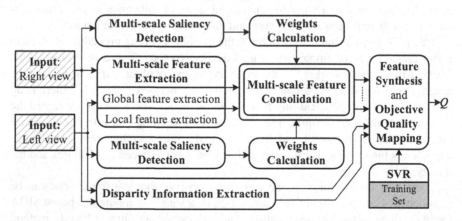

Fig. 7.12 Framework of the no-reference SIQA method proposed in (Xu et al. 2017)

subtracted contrast normalized (MSCN) coefficients are computed. The MSCN coefficients are proved a good reflection of image distortions (Mittal et al. 2012b). More specifically, the contribution of the MSCN coefficients often appears as Gaussian-like. Thus, by modeling the distribution using a generalized Gaussian distribution (GGD) model, the parameters of the model give away the distortion contained in an image (Zhou et al. 2016). The GGD model after zero mean subtraction is represented as

$$f\left(x|\alpha,\sigma^2\right) = \frac{\alpha\sqrt{\Gamma\left(\frac{3}{\alpha}\right)}}{2\sigma\sqrt{\Gamma\left(\frac{1}{\alpha}\right)^3}} \exp\left(-\left(\frac{|x|\sqrt{\Gamma\left(\frac{3}{\alpha}\right)}}{\sigma\sqrt{\Gamma\left(\frac{1}{\alpha}\right)^3}}\right)^\alpha\right) \quad (7.80)$$

where x is the MSCN coefficient and Γ denotes the Gamma function. The parameters employed as the features are the variations σ^2 and the shape indicator α. The global feature $\mathbf{f}_g$ is composed by them,

$$\mathbf{f}_g = \left\{\alpha, \sigma^2\right\} \quad (7.81)$$

To take the multiscale property into account, an additional scale is manually constructed by a simple pyramid decomposition. Therefore, from the two scales, we are actually obtaining a feature vector with size 4,

$$\mathbf{f}_g = \left\{\alpha_1, \sigma_1^2, \alpha_2, \sigma_2^2\right\} \quad (7.82)$$

where the subscript represents the scale index.

The local features are captured by computing local entropy, which is another image feature proved quality-aware (Liu et al. 2014). There exist different definitions for the concept of entropy, and the authors adopt two of them, the spatial and spectral entropies. As it is desired to extract features in a local way, the images are firstly divided into blocks. The spatial entropy (E_s) represents the probability distribution of the local pixels, which is a statistic in pixel-level,

$$E_s(x) = -\frac{\sum_x \left(N\left(\log_2 N - \log_2 \sum_x N\right)\right)}{\sum_x N} \quad (7.83)$$

where N is the pixel number within a block.

The spectral entropy E_f is defined as the probability distribution of DCT coefficients within a local block, which is also a good indicator of distortion type,

$$E_f(x') = -\frac{\sum_x \left(x'^2\left(\log_2(x'^2) - \log_2 \sum_x x'^2\right)\right)}{\sum_x x'^2} \quad (7.84)$$

where x' denotes the DCT coefficients of pixel x.

The mean and skew of the entropy metrics (both spatial and spectral) are adopted as the local features, to try to eliminate the dependency on image content. The features are captured from the three scales, also constructed using pyramid decomposition. Moreover, only the central 60% of data are employed to derive the mean and skew. As a result, the local feature vector is of size 12,

$$\mathbf{f}_l = \left\{ \begin{array}{l} \text{mean}(E_{s1}), \text{skew}(E_{s1}), \text{mean}(E_{f1}), \text{skew}(E_{f1}), \\ \text{mean}(E_{s2}), \text{skew}(E_{s2}), \text{mean}(E_{f2}), \text{skew}(E_{f2}), \\ \text{mean}(E_{s3}), \text{skew}(E_{s3}), \text{mean}(E_{f3}), \text{skew}(E_{f3}) \end{array} \right\} \tag{7.85}$$

Therefore, together there are 16 quality indices for either the left or the right view.

The features are extracted from both left and right views. To synthesize the features from the two sources as well as to make the indices more accurate, visual saliency is made use of, which is also computed from the multiple corresponding scales. On a certain scale, denoting $\mathbf{f}_L$ and $\mathbf{f}_R$ as the features from the left and right views, and $\mathbf{S}_L$ and $\mathbf{S}_R$ as the saliency maps for the left and right views (the Frobenius norms of $\mathbf{S}_L$ and $\mathbf{S}_R$ are denoted as $\|\mathbf{S}_L\|$ and $\|\mathbf{S}_R\|$), the final feature vector $\mathbf{f}$ is

$$\mathbf{f} = \frac{\|\mathbf{S}_L\|}{\|\mathbf{S}_L\| + \|\mathbf{S}_R\|} \mathbf{f}_L + \frac{\|\mathbf{S}_R\|}{\|\mathbf{S}_L\| + \|\mathbf{S}_R\|} \mathbf{f}_R \tag{7.86}$$

By far, the method has been following the typical framework of the weight computation-based SIQA methods. One evident innovation in the work is that the disparity maps are also employed to produce extra quality-aware features. As we have previously talked about, a traditional way to make use of the disparity maps is to find the difference between the disparity maps of the reference and distorted image pairs (Benoit et al. 2008b). However, the proposed method is a no-reference one, making the strategy unavailable.

The employed scheme is to capture the statistical changes of the disparity maps, which is operated in a quite innovative way. If the ideal disparity maps follow certain kind of distribution, then the distortions will make the distribution not as uniform. Specifically, the kurtosis (k) and skew (s) of a disparity map $\mathbf{D}$ are computed,

$$k = \frac{\frac{1}{N}\sum \left(\mathbf{D} - \frac{1}{N}\sum \mathbf{D}\right)^4}{\frac{1}{N}\sum \left(\mathbf{D} - \frac{1}{N}\sum \mathbf{D}\right)^2} \tag{7.87}$$

$$s = \frac{\frac{1}{N}\sum \left(\mathbf{D} - \frac{1}{N}\sum \mathbf{D}\right)^3}{\left(\frac{1}{N}\sum \left(\mathbf{D} - \frac{1}{N}\sum \mathbf{D}\right)^2\right)^{\frac{3}{2}}} \tag{7.88}$$

The disparity map $\mathbf{D}$ is derived using the SSIM-based estimation scheme (Chen et al. 2013b).

Including the two indices computed from the disparity maps, there are 18 quality indices for a stereoscopic image. Then, support vector machine is adopted to map the indices into the final quality score.

7.5.5 Compact Framework Considering Visual Discomfort and Binocular Rivalry

The last example we would like to introduce is another no-reference method, proposed in (Ding and Zhao 2017). Similar to the method proposed in (Xu et al. 2017), the authors hold a critical attitude toward the thought of constructing cyclopean images for its time consuming. Moreover, there are at least three innovations in their work. Firstly, the visual discomfort when viewing stereo scenes (see Sect. 7.2) is taken into account. Secondly, the method adopts a framework that makes use of neither cyclopean images nor weight computation. Thirdly, attempts are made to make the flow of the method as compact as possible so that a parallel framework can be effective.

The framework of the method is given in Fig. 7.13. As is shown, there are three main paths of image processing, which are organized in parallel. In the upper path, the images are segmented adaptively, the main purpose of which is to approximate visual discomfort. The middle path is the computation of image local energy. The local energy is considered both a quality-aware feature (Ding et al. 2017) and a good reflection of visual attention when computing binocular rivalry (Chen et al. 2013b). Finally, in the lower path, visual saliency is computed for assistance.

Let's begin the discussion from the upper path. The adaptive image segmentation is completed using simple linear iterative clustering (SLIC). Because each image pixel can be represented by five values, the two-dimensional spatial coordinates and the three color-channels (R, G, and B, in this case), we can regard a

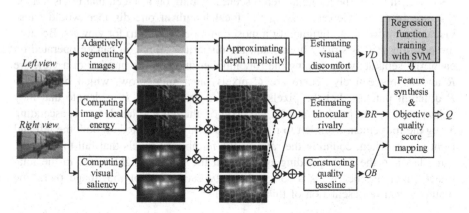

Fig. 7.13 Framework of the no-reference SIQA method proposed in (Ding and Zhao 2017)

pixel as a point in a five-dimensional hyperspace. A linear clustering method, K-means, is adopted, which iteratively reduces the cost function

$$J = \sum_{n=1}^{N} \sum_{k=1}^{K} r_{nk} \|x_n - \mu_k\|^2 \qquad (7.89)$$

where x_n denotes the n-th pixel among the N pixels, μ_k denotes the center of the k-th cluster among the K clusters, and r_{nk} is 1 when x_n belongs to the k-th cluster and 0 if otherwise.

The super-parameter K decides how many clusters the data are expected to be classified into, which is also how many patches that are supposed to be obtained during the image segmentation. By SLIC, the images are segmented in a manner that the similar pixels in an irregular region should belong to the same patch. Considering an ideal occasion that the pixels of one object are classified into a single patch. Because the object should appear in different spatial positions in the left and right views (a horizonal position shift, as we know), how large the position shift is makes a rough disparity estimation of the object. Since objects may appear in different sizes, multiple scales are constructed by choosing K wisely. Specifically, four scales are constructed, by defining K as

$$K(s)|_{s=0,1,2,3} = \left\lfloor \frac{N}{\alpha \cdot \beta^s} \right\rfloor \qquad (7.90)$$

Obviously, the values of K are selected from a sequence of common ratio β, which is empirically set as 4. The parameter α defines the average of the patches at scale 0, empirically set as 64. Setting β as 4 determines that the different scales are constructed using a sampling ratio of 2×2, which is basically reasonable in the construction of multiple image scales.

The measurement of visual discomfort is closely connected with the distribution of the depth information. The reason behind this fact has been introduced in Sect. 7.2. In short, the artificial stereo scenes display on a screen that holds a fixed distance from the viewers. Fixing the focal length at one distance would cause visual discomfort (e.g., fatigue, for a most common symptom) for viewers. Because the position shift is strictly horizontal, the disparity estimation can be operated on each row separately. Assuming the row number and column number of a view are R and C, respectively. There are C pixels on a single row, which belongs to P different patches. Let the pixels be labeled as 1, 2, ..., P to denote that they belong to different patches, and let the pixel on the left end labeled as 1, ascending along the movement from left to right. The labeling is operated on both the left and right views. Then, counting the ratio of the number of pixels that fail to have the same label in the corresponding position on the left and right views to the total image pixel number gives a rough disparity estimation of the image pair. The mathematical representation of this process is

$$VD = \frac{1}{R \cdot C} \sum_{r=1}^{R} \sum_{c=1}^{C} \mathrm{sgn}(l_L(r,c), l_R(r,c)) \qquad (7.91)$$

where VD is the estimated degree of visual discomfort, $l_L(r, c)$ denotes the label of the pixel at position (r, c) from the left view, and $l_R(r, c)$ has similar meaning. The signal function sgn is defined as

$$\mathrm{sgn}(x,y) = \begin{cases} 1, & x = y \\ 0, & x \neq y \end{cases} \qquad (7.92)$$

Different from conventional visual discomfort estimation methods that usually contain two serial stages, i.e., computing disparity maps and extracting features from the maps, the novel method to estimate visual discomfort does not contain an explicit disparity computation. Therefore, the estimation operates quite efficiently. On the other hand, the multiscale approach allows the accuracy to be largely preserved. Moreover, the segmentation results are used in the processing of other parallel paths. This data re-utilization makes the flow more compact and efficient.

The local energy is defined as the magnitude of frequency responses. As a good simulation of HVS responses, log Gabor filter is employed. The response of log Gabor filter inputted by pixels in polar form can be represented as

$$\mathbf{LG}(f,\theta) = \exp\left(-\frac{\left(\log\frac{f}{f_0}\right)^2}{2\left(\log\frac{\sigma_f}{f_0}\right)^2} - \frac{(\theta - \theta_0)^2}{2\sigma_\theta^2}\right) \qquad (7.93)$$

where θ_0 and f_0 are the orientation and central frequency of the filter, respectively; σ_f and σ_θ determine the scale and angular bandwidths, respectively. By modifying the parameters, the filter is defined with certain scales and orientations. To be consistent with the segmentation, four scales are constructed; and four orientations $(0°, 45°, 90°$ and $135°)$ are adopted, as a conventional usage.

The magnitude of $\mathbf{LG}$ serves as the local energy. In addition, on each scale, the responses from the four orientations are summed up on each scale, for the convenience of computation, the energy map $\mathbf{EM}$ is

$$\mathbf{EM} = \sum_{i=0}^{3} \sqrt{(\mathrm{Re}(\mathbf{LG}_{i \times 45°}))^2 + (\mathrm{Im}(\mathbf{LG}_{i \times 45°}))^2} \qquad (7.94)$$

where the subscripts denote the orientation of the filter.

Moreover, the local energy is estimated on block-level rather than pixel-level, again for the sake of computational simplicity. The local energy within the adaptively segmented irregular patches is taken as one value, by computing the mean of the $\mathbf{EM}$ within the patch. For a certain scale where the images are divided into

K patches, we have K numbers to denote the local energy. To make the results more robust, only γ $(0 < \gamma < 1)$ of the K values are retained to represent the energy level of the whole image. The selection is according to visual saliency, computed using the gradient-based visual saliency (GBVS) (Hou et al. 2012). The parameter γ is empirically set at 70%. The median of the 70% retained values is considered the energy level for a whole view.

For a scale, let E_L and E_R denote the local energy levels of the left and right views, respectively, then the degree of binocular rivalry (BR) can be approximated by

$$BR = \max\left\{\frac{E_L}{E_R}, \frac{E_R}{E_L}\right\} \tag{7.95}$$

By far, the degrees of two major binocular effects have been computed, i.e., VD and BR. Yet there has not been an index that is directly related to image quality. Therefore, a quality baseline (QB) is created to make a comprehensive description of image quality,

$$QR = \frac{E_L + E_R}{2} \tag{7.96}$$

Finally, the indices produced by VD, BR, and QB are mapped to the final image quality score using support vector machine.

In the last of this Section, we would like to recommend readers some other literatures with newly proposed SIQA methods in recent years: (Shao et al. 2013; Zhou et al. 2014a, b; Su et al. 2015; Shao et al. 2015a; Qi et al. 2015; Li et al. 2015; Md et al. 2015; Zhou and Yu 2016; Shao et al. 2016a, b; Appina et al. 2016; Liu et al. 2016; Wang et al. 2016; Lv et al. 2016; Zhang et al. 2016; Hachicha et al. 2017; Geng et al. 2017; Oh et al. 2017; Zhou et al. 2017a, b). We are aware that this is definitely an incompletely collection of worth reading works for you, so if you have interests in this research field, it is highly encouraged to go online and search for more related works.

7.6 Summary and Future Work

Obviously, significant progresses are made in both subjective and objective SIQA study in the last decade. To summarize them, as well as to offer some prospects for the future, we will give some further analysis and related discussions, as follows.

In subjective SIQA, multiple databases have been constructed in recent years. From 2D IQA, we learn that it is important to include various distortion types in a database, so that the robustness of objective methods can be validated. But in SIQA, other issues shall be taken into account. A major issue is the introduction of asymmetrical distortion, a reliable approach for testing whether an SIQA method is really "stereoscopic." Fortunately, because the binocular rivalry phenomenon is noticed since very early research stage, the asymmetric distortion is well taken care of in many established databases (Sazzad et al. 2009; Wang et al. 2009; Chen et al. 2013a, b; Wang et al. 2015). Some other issues are pointed out in (Wang et al. 2015). For one thing, it would be also important to carry out experiments to label the subjective quality scores for the monocular views of stereo image pairs in stereoscopic databases, besides the scores for the stereoscopic images. For another, mixed distortions should be applied on reference and tested images. These issues, either old problems similar to what we have seen in 2D IQA or new ones occurred due to the properties of stereoscopic images, demand the database builders to account for so that the databases are sufficient enough to evaluate both the accuracy and robustness of objective methods.

However, current SIQA databases are mostly too small in size, especially comparing to existing 2D databases. Concretely, most SIQA databases contain less than 1000 images, but the largest-scale 2D IQA database includes around 3000 images. The relative small database scale makes the results less persuasive and forbids machine learning-based methods to obtain ideal models. Another problem partly corresponding to the size problem is the distortion types contained; existing databases usually include only 3–5 distortion types at most. The distortions that are common for stereoscopic images are likely to differ with those in 2D field due to the variation of application scenarios. For instance, because 3D video is a popular application, the distortion that video coding brings about should be taken care of, such as H.264 in NBU database (Wang et al. 2009). Generally, researchers have to consider more issues and operate under more complex procedures when building an SIQA subjective database than its 2D counterpart, and there is still room for improvement.

As for objective methods, the main issue is how to combine 2D methods and stereo algorithms to make them applicable for stereoscopic images. On one hand, 2D IQA have provided extensive knowledge about how human perceives image quality and how to model this quality perception mathematically and objectively. For stereoscopic images, this kind of knowledge is certainly useful as well. The tool used to extract quality-aware features from 2D images can be introduced for monocular views and cyclopean images quite straightforwardly. For monocular views and cyclopean images, novel ways for feature extraction seem more likely to be innovations for 2D IQA, but there might be slight difference from 2D images, which is worth exploring in the future. For other objects that we can extract features from, such as the depth or disparity maps, however, the traditional 2D quality-aware features are evidently not applicable, so how these features help for SIQA is also worth studying.

On the other hand, a more inherent problem for SIQA is how to extend these 2D methods into 3D field. In many previous works, researchers have tried to make use of the depth information, the weights to combine quality scores from the left and right monocular views, and the cyclopean images constructed with the help of binocular disparity, and observed progresses over time. It is exactly the difference in these extending schemes that allow us to extract features from various sources for a comprehensive description of image quality. Unfortunately, due to our shallow understanding upon human binocular vision, there are many unsettled problems causing troubles for applying the extension. For example, how disparity maps are affecting quality perception and how cyclopean images are generated in our minds are all unsolved puzzles. So, related discoveries about human binocular vision and binocular effects are highly demanding.

Another trend of objective SIQA study is the combination with machine learning techniques, which can be helpful in a large variety of ways. To name a few, regression function training can build a projection that maps evaluated quality from multiple sources to one final score; dictionary learning allows us to make use of discrete and even enumerating features more conveniently and efficiently; convolutional neural networks can even automatically extract features and continuously optimize the feature extraction process. From a present view, combinational feature extraction and pooling schemes and the introduction of machine learning techniques are likely to offer satisfactory performance and breakthroughs for objective SIQA research in the near future.

References

Appina, B., Khan, S., & Channappayya, S. S. (2016). No-reference stereoscopic image quality assessment using natural scene statistics. *Signal Processing: Image Communication, 43,* 1–14.

Barten, P. G. J. (2004). Formula for the contrast sensitivity of the human eye. *Proceedings of SPIE, 5294,* 231–238.

Bengio, Y. (2009). *Learning deep architectures for AI.* Boston, MA, USA: New Publishers Inc.

Benoit, A., Le Callet, P., Campisi, P., & Cousseau, R. (2008a). Quality assessment of stereoscopic images. *EURASIP Journal on Image and Video Processing, 2008,* 659024.

Benoit, A., Le Callet, P., Campisi P., & Cousseau, R. (2008b). Using disparity for quality assessment of stereoscopic images. In *IEEE International Conference on Image Processing,* October 2008, San Diego, United States.

Bensalma, R., & Larabi, M. C. (2013). A perceptual metric for stereoscopic image quality assessment based on the binocular energy. *Multidimensional Systems and Signal Processing, 24*(2), 281–316.

Blake, R. (1977). Threshold conditions for binocular rivalry. *Journal of Experimental Psychology: Human Perception and Performance, 3*(2), 251–257.

Blake, R., & Fox, R. (1973). Psychophysical inquiry into binocular summation. *Perception and Psychophysics, 14*(1), 161–185.

Campisi, P., Le Callet, P., & Marini, E. (2007). Stereoscopic image quality assessment. In *European Signal Processing Conference,* September 2007, Poznan, Poland.

Carnec, M., Le Callet, P., & Barba, D. (2003). An image quality assessment based on the perception of structural information. *Proceedings of International Conference on Image Processing, 3,* 185–188.

Chandler, D. M., & Hemami, S. S. (2007). VSNR: a wavelet-based visual signal-to-noise ratio for natural images. *IEEE Transactions on Image Processing, 16*(9), 2284–2298.

Chen, J., Zhou, J., Sun, J., & Bovik, A. C. (2017). Visual discomfort prediction on stereoscopic images without explicit disparities. *Signal Processing: Image Communication, 51,* 50–60.

Chen, M.-J., Cormack, L. K., & Bovik, A. C. (2013a). No-reference quality assessment of natural stereopairs. *IEEE Transactions on Image Processing, 22*(9), 3379–3397.

Chen, M.-J., Su, C.-C., Kwon, D.-K., Cormack, L. K., & Bovik, A. C. (2013b). Full-reference quality assessment of stereopairs accounting for rivalry. *Signal Processing: Image Communication, 28,* 1143–1155.

Choi, J., Kim, D., Choi, S., & Sohn, K. (2012). Visual fatigue modeling and analysis for stereoscopic video. *Optical Engineering, 51*(1), 017206.

Ding, J., & Sperling, G. (2006). A gain-control theory of binocular combination. *Proceedings of the National Academy of Sciences of the United States of America, 103*(4), 1141–1146.

Ding, Y., & Zhao, Y. (2017a). No-reference quality assessment for stereoscopic images considering visual discomfort and binocular rivalry. *Electronics Letters, 53*(25), 1646–1647.

Ding, Y., Zhao, Y., & Zhao, X. (2017b). Image quality assessment based on multi-feature extraction and synthesis with support vector regression. *Signal Processing: Image Communication, 54,* 81–92.

Ehrenstein, W. H., Arnold-Schulz-Gahmen, B. E., & Jaschinski, W. (2005). Eye preference within the context of binocular function. *Graefe's Archive for Clinical and Experimental Ophthalmology, 243*(9), 926–932.

Eser, I., Durrie, D. S., Schwendeman, F., & Stahl, J. E. (2008). Association between ocular dominance and refraction. *Journal of Refractive Surgery, 24*(7), 685–689.

Fahle, M. (1982). Binocular rivalry: Suppression depends on orientation and spatial frequency. *Vision Research, 22*(7), 787–800.

Fahle, M. (1987). Two eyes, for what? *Naturwissenchaften, 74*(8), 383–385.

Felzenszwalb, P. F., & Huttenlocher, D. P. (2006). Efficient belief propagation for early vision. *International Journal of Computer Vision, 70*(1), 41–54.

Geng, X., Shen, L., Li, K., & An, P. (2017). A stereoscopic image quality assessment model based on independent component analysis and binocular fusion property. *Signal Processing: Image Communication, 52,* 54–63.

Ghadiyaram, D., & Bovik, A. C. (2014). Blind image quality assessment on real distorted images using deep belief nets. *Proceedings of IEEE Global Conference on Signal and Information Processing, 2014,* 946–950.

Goldmann, L., De Simone, F., & Ebrahimi, T. (2010). Impact of acquisition distortions on the quality of stereoscopic images. In *International Workshop on Video Processing, Quality Metrics, and Consumer Electronics*, Janaury 2010, Scottsdale, United States.

Gu, K., Zhai, G., Yang, X., & Zhang, W. (2014). Deep learning network for blind image quality assessment. *Proceedings of IEEE International Conference on Image Processing, 2014,* 511–515.

Hachicha, W., Kaaniche, M., Beghdadi, A., & Cheikh, F. A. (2017). No-reference stereo image quality assessment based on joint wavelet decomposition and statistical models. *Signal Processing: Image Communication, 54,* 107–117.

He, L., Tao, D., Li, X., & Gao, X. (2012). Sparse representation for blind image quality assessment. In *IEEE International Conference on Computer Vision and Pattern Recognition* (pp. 1146–1153).

Hinton, G. E., Osindero, S., & Teh, Y. W. (2006). A fast learning algorithm for deep belief nets. *Neural Computing, 18*(7), 1527–1554.

Hirschmuller, H. (2005). Accurate and efficient stereo processing by semi-global matching and mutual information. In *International Conference on Computer Vision and Pattern Recognition*, June 2005, San Diego, United States.

Howard, I. P., & Roger, B. J. (1995). *Binocular vision and stereopsis.* New York: Oxford University Press.

Hou, W., Gao, X., Tao, D., & Li, X. (2015). Blind image quality assessment via deep learning. *IEEE Transactions on Neural Network and Learning Systems, 26*(6), 1275–1286.

Hou, X., Harel, J., & Koch, C. (2012). Image signature: Highlighting sparse salient regions. *IEEE Transactions on Pattern Analysis and Machine Intelligence, 34*(1), 194–201.

Huang, C. B., Zhou, J., Lu, Z. L., & Zhou, Y. (2011). Deficient binocular combination reveals mechanisms of anisometropic amblyopia: Signal attenuation and interocular inhibition. *Journal of Vision, 11*(6), 4.

Huang, C. B., Zhou, J., Zhou, Y., & Lu, Z. L. (2010). Contrast and phase combination in binocular vision. *PLoS ONE, 5*(12), e15075.

Julesz, B. (1960). Binocular depth perception of computer-generated patterns. *Bell System Technical Journal, 39*(5), 1125–1162.

Julesz, B. (1971). *Foundations of cyclopean perception*. Chicago: The University of Chicago Press.

Kang, L., Ye, P., Li, Y., & Doermann, D. (2014). Convolutional neural networks for no-reference image quality assessment. *Proceedings of IEEE Conference on Computer Vision and Pattern Recognition, 2014*, 1733–1740.

Kaptein, R. G., Kuijstersa, A., Lambooij, M. T. M., Ijsselsteijn, W. A., & Heynderickx, I. (2008). Performance evaluation of 3D-TV systems. *Proceedings of SPIE, 6808*, 680819.

Kaufman, L. (1974). *Sight and mind: An introduction to visual perception*. London, U.K.: Oxford Press.

Khan, A. Z., & Crawford, J. D. (2001). Ocular dominance reverses as a function of horizontal gaze angle. *Vision Research, 41*(14), 1743–1748.

Kim, D., & Sohn, K. (2011). Visual fatigue prediction for stereoscopic image. *IEEE Transactions on Circuits and Systems for Video Technology, 21*(2), 231–236.

Klaus, A., Sormann, M., & Karner, K. (2006). Segment-based stereo matching using belief propagation and a self-adapting dissimilarity measure. *Proceedings of the 18th International Conference on Pattern Recognition, 3*, 15–18.

Konolige, K. (1997). Small vision system: Hardware and implementation. In *Proceedings of the 8th International Symposium in Robotic Research* (pp. 203–212).

Lambooij, M. T. M., Ijsselsteijn, W. A., & Heynderickx, I. (2009). Visual discomfort in stereoscopic displays: A review. In *Conference on Stereoscopic Displays and Virtual Reality Systems XIV*, Janauary 2009, San Jose, United States.

Larson, E. C., & Chandler, D. M. (2010). Most apparent distortion: Full-reference image quality assessment and the role of strategy. *Journal of Electronic Imaging, 19*(1), 1–21.

Levelt, W. J. M. (1965). Binocular brightness averaging and contour information. *British Journal of Psychology, 56*(1), 1–13.

Levelt, W. J. M. (1966). The alternation process in binocular rivalry. *British Journal of Psychology, 57*(3–4), 225–238.

Levelt, W. J. M. (1968). *On binocular rivalry*. Paris: Mouton.

Li, K., Shao, F., Jiang, G., & Yu, M. (2015). Joint structure-texture sparse coding for quality prediction of stereoscopic images. *Electronics Letters, 51*(24), 1994–1995.

Lin, Y. H., & Wu, J. L. (2014a). Quality assessment of stereoscopic 3D image compression by binocular integration behaviors. *IEEE Transactions on Image Processing, 23*(4), 1527–1542.

Liu, L., Liu, B., & Huang, H. (2014b). No-reference image quality assessment based on spatial and spectral entropies. *Signal Processing: Image Communication, 29*(8), 856–863.

Liu, Y., Yang, J., Meng, Q., Lv, Z., Song, Z., & Gao, Z. (2016). Stereoscopic image quality assessment method based on binocular combination saliency model. *Signal Processing, 125*, 237–248.

Lv, Y., Yu, M., Jiang, G., Shao, F., Peng, Z., & Chen, F. (2016). No-reference stereoscopic image quality assessment using binocular self-similarity and deep neural network. *Signal Processing: Image Communication, 47*, 349–357.

Md, S. K., Appina, B., & Channappayya, S. S. (2015). Full-reference stereo image quality assessment using natural scene statistics. *IEEE Signal Processing Letters, 22*(11), 1985–1989.

Meegan, D. V., Stelmach, L. B., & Tam, W. J. (2001). Unequal weighting of monocular inputs in binocular combination: Implications for the compression of stereoscopic imagery. *Journal of Experimental Psychology: Applied, 7*(2), 143–153.

Mittal, A., Moorthy, A. K., & Bovik, A. C. (2012a). Making image quality assessment robust. In *Conference Record of the Asilomar Conference on Signals, Systems and Computers* (pp. 1718–1722).

Mittal, A., Moorthy, A. K., & Bovik, A. C. (2012b). No-reference image quality assessment in the spatial domain. *IEEE Transactions on Image Processing, 21*(12), 4695–4708.

Mohamed, A. R., Sainath, T. N., Dahl, G., Ramabhadran, B., Hinton, G. E. & Picheny, M. A. (2011). Deep belief networks using discriminative features for phone recognition. In *IEEE International Conference on Acoustics, Speech, and Signal Processing* (pp. 5060–5063).

Moorthy, A. K., Su, C.-C., Mittal, A., & Bovik, A. C. (2013). Subjective evaluation of stereoscopic image quality. *Signal Processing: Image Communication, 28*(8), 870–883.

Nojiri, Y., Yamanoue, H., Hanazato, A., & Okano, F. (2003). Measurement of parallax distribution and its application to the analysis of visual comfort for stereoscopic HDTV. *Proceedings of SPIE, 5006,* 195–205.

Oh, H., Ahn, S., Kim, J., & Lee, S. (2017). Blind deep S3D image quality evaluation via local to global feature aggregation. *IEEE Transactions on Image Processing, 26*(10), 4923–4936.

Ojala, T., Pietikäinen, M., & Mäenpää, T. (2002). Multiresolution gray-scale and rotation invariant texture classification with local binary patterns. *IEEE Transactions on Pattern Recognition and Machine Intelligence, 24*(7), 971–987.

Park, J., Lee, S., & Bovik, A. C. (2014). 3D visual discomfort prediction: vergence, foveation, and the physiological optics of accommodation. *IEEE Journal of Selected Topic in Signal Processing, 8*(3), 415–427.

Porac, C., & Coren, S. (1975). Is eye dominance a part of generalized laterality? *Perceptual and Motor Skills, 40*(3), 763–769.

Qi, F., Zhao, D., & Gao, W. (2015). Reduced reference stereoscopic image quality assessment based on binocular perceptual information. *IEEE Transactions on Multimedia, 17*(12), 2338–2344.

Reiss, M. R. (1997). Ocular dominance: Some family data. *Laterality, 2*(1), 7–16.

Richardt, C., Świrski, L., Davies, I. P., & Dodgson, N. A. (2011). Predicting stereoscopic viewing comfort using a coherence-based computational model. In *Proceedings of the International Symposium on Computational Aesthetics in Graphics, Visualization, and Imaging* (pp. 97–104).

Saad, M. A., Bovik, A. C., & Charrier, C. (2012). Blind image quality assessment: A natural scene statistics approach in the DCT domain. *IEEE Transactions on Image Processing, 21*(8), 3339–3352.

Sazzad, Z. M. P., Yamanaka, S., Kawayokeita, Y., & Horita, Y. (2009). Stereoscopic image quality prediction. In *International Workshop on Quality of Multimedia Experience*, July 2009, San Diego, United States.

Scharstein, D., & Szeliski, R. (2002). A taxonomy and evaluation of dense two-frame stereo correspondence algorithms. *International Journal of Computer Vision, 47,* 7–42.

Seuntiëns, P. J. H., Meesters, L. M. J., & Ijsselsteijn, W. A. (2006). Perceived quality of compressed stereoscopic images: Effects of symmetric and asymmetric JPEG coding and camera separation. *ACM Transactions on Applied Perception, 3,* 95–109.

Shao, F., Lin, W., Gu, S., Jiang, G., & Srikanthan, T. (2013). Perceptual full-reference quality assessment of stereoscopic images by considering binocular visual characteristics. *IEEE Transactions on Image Processing, 22*(5), 1940–1953.

Shao, F., Li, K., Lin, W., Jiang, G., & Yu, M. (2015a). Using binocular feature combination for blind quality assessment of stereoscopic images. *IEEE Signal Processing Letters, 22*(10), 1548–1551.

Shao, F., Li, K., Lin, W., Jiang, G., Yu, M., & Dai, Q. (2015b). Full-reference quality assessment of stereoscopic images by learning binocular receptive field properties. *IEEE Transactions on Image Processing, 24*(10), 2971–2983.

Shao, F., Li, K., Lin, W., Jiang, G., & Dai, Q. (2016a). Learning blind quality evaluator for stereoscopic images using joint sparse representation. *IEEE Transactions on Multimedia, 18*(10), 2104–2114.

Shao, F., Tian, W., Lin, W., Jiang, G., & Dai, Q. (2016b). Toward a blind deep quality evaluator for stereoscopic images based on monocular and binocular interactions. *IEEE Transactions on Image Processing, 25*(5), 2059–2074.

Sheikh, H. R., Sabir, M. F., & Bovik, A. C. (2006). A statistical evaluation of recent full reference image quality assessment algorithms. *IEEE Transactions on Image Processing, 15*(11), 3440–3451.

Su, C. C., Bovik, A. C., & Cormack, L. K. (2011). Natural scene statistics of color and range. In *IEEE International Conference on Image Processing* (pp. 257–260).

Su, C. C., Cormack, L. K., & Bovik, A. C. (2015). Oriented correlation models of distorted natural images with application to natural stereopair quality evaluation. *IEEE Transactions on Image Processing, 24*(5), 1685–1699.

Tang, H., Niranjan, J., & Ajay, K. (2014). Blind image quality assessment using semi-supervised rectifier networks. *Proceedings of IEEE Conference on Computer Vision and Pattern Recognition, 2014,* 2877–2884.

Wang, H., Yang, J., Lu, W., Li, B., Tian, K., & Meng, Q. (2016). No-reference stereoscopic image-quality metric accounting for left and right similarity map and spatial structure degradation. *Optics Letters, 41*(24), 5640–5643.

Wang, X., Yu, M., Yang, Y., & Jiang, G. (2009). Research on subjective stereoscopic image quality assessment. *Proceedings of SPIE, 7255,* 725509.

Wang, J., & Wang, Z. (2014). Perceptual quality of asymmetrically distorted stereoscopic images: the role of image distortion types. In *International Workshop on Video Processing & Quality Metrics for Consumer Electronics*, Chandler, Arizona, USA.

Wang, J., Rehman, A., Zeng, K., Wang, S., & Wang, Z. (2015). Quality prediction of asymmetrically distorted stereoscopic 3D images. *IEEE Transactions on Image Processing, 24*(11), 3400–3414.

Wang, J., Zeng, K., & Wang, Z. (2014). Quality prediction of asymmetrically distorted stereoscopic images from single views. In *IEEE International Conference on Multimedia and Expo Workshops*, July 14–18, Chengdu, China.

Wang, Z., & Bovik, A. C. (2002). A universal image quality index. *IEEE Signal Processing Letters, 9*(3), 81–84.

Wang, Z., & Li, Q. (2011). Image content weighting for perceptual image quality assessment. *IEEE Transactions on Image Processing, 20*(5), 1185–1198.

Wang, Z., & Shang, X. (2006). Spatial pooling strategies for perceptual image quality assessment. In *Proceeding of IEEE International Conference on Image Processing* (pp. 2945–2948).

Wang, Z., & Simoncelli, E. P. (2005). Reduced-reference image quality assessment using a wavelet-domain natural image statistic model. *Human Vision and Electronic Imaging X, Proceedings of the SPIE, 5666,* 149–159.

Wang, Z., Simoncelli, E. P., & Bovik, A. C. (2003). Multi-scale structural similarity for image quality assessment. In: *Conference Records of the Asilomar Conference on Signals, Systems and Computers* (pp. 1398–1402).

Wang, Z., Bovik, A. C., Sheikh, H. R., & Simoncelli, E. P. (2004). Image quality assessment: from error visibility to structural similarity. *IEEE Transactions on Image Processing, 13*(4), 600–612.

Wu, J., Lin, W., Shi, G., Zhang, Y., Dong, W., & Chen, Z. (2015). Visual orientation selectivity based structure description. *IEEE Transactions on Image Processing, 24*(11), 4602–4613.

Xu, X., Zhao, Y., & Ding, Y. (2017). No-reference stereoscopic image quality assessment based on saliency-guided binocular feature consolidation. Electronics Letter, *53*(22): 1468–1470.

Xue, W., Mou, X., Zhang, L., Bovik, A. C., & Feng, X. (2014). Blind image quality assessment using joint statistics of gradient magnitude and Laplacian features. *IEEE Transactions on Image Processing, 23*(11), 4850–4862.

Yamashita, T., Tanaka, M., Yoshida, E., Yamauchi, Y., & Fujiyoshii, H. (2014). To be Bernoulli or to be Gaussian, for a restricted Boltzmann machine. In: *Proceedings of International Conference on Pattern Recognition* (pp. 1520–1525).

Yang, J., Hou, C., & Xu, R. (2010). New metric for stereo image quality assessment based on HVS. *International Journal of Image Systems and Technology, 20*(4), 2010.

Yang, J., Hou, C., Zhou, Y., Zhang, Z., & Guo, J. (2009). Objective quality assessment method of stereo images. In *2009 3DTV-Conference: The True Vision—Capture, Transmission and Display of 3D Video*, May 2009, Potsdam, Germany.

Yano, S., Ide, S., Mitsuhashi, T., & Thwaites, H. (2002). A study of visual fatigue and visual discomfort for 3D HDTV/HDTV images. *Displays, 23*(4), 191–201.

Yang, Y., & Blake, R. (1991). Spatial frequency tuning of human stereopsis. *Vision Research, 31*(7–8), 1176–1189.

Yasakethu, S. L. P., Hewage, C. T. E. R., Fernando, W. A. C., & Kondoz, A. M. (2008). Quality analysis for 3D video using 2D video quality models. *IEEE Transactions on Consumer Electronics, 54*(4), 1969–1976.

You, J., Xing, L., Perkis, A., & Wang, X. (2010). Considering binocular spatial sensitivity in stereoscopic image quality assessment. In: *Proceedings of the International Workshop of Video Processing* (pp. 61–66).

Zhang, W., Qu, C., Ma, L., Guan, J., & Huang, R. (2016). Learning structure of stereoscopic image for no-reference quality assessment with convolutional neural network. *Pattern Recognition, 59*, 176–187.

Zhang, Y., & Chandler, D. M. (2015). 3D-MAD: A full reference stereoscopic image quality estimator based on binocular lightness and contrast perception. *IEEE Transactions on Image Processing, 24*(11), 3810–3825.

Zhao, Y., Ding, Y., & Zhao, X. (2016). Image quality assessment based on complementary local feature extraction and quantification. *Electronics Letters, 52*(22), 1849–1851.

Zhou, J., Jiang, G., Mao, X., Yu, M., Shao, F., Peng, Z. & Zhang, Y. (2011). In *IEEE International Conference on Visual Communications and Image Processing*, November 2011, Tainan, Taiwan.

Zhou, W., Jiang, G., Yu, M., Shao, F., & Peng, Z. (2014a). PMFS: A perceptual modulated feature similarity metric for stereoscopic image quality assessment. *IEEE Signal Processing Letters, 21*(8), 1003–1006.

Zhou, W., Jiang, G., Yu, M., Shao, F., & Peng, Z. (2014b). Reduced-reference stereoscopic image quality assessment based on view and disparity zero-watermarks. *Signal Processing: Image Communication, 29*(1), 167–176.

Zhou, W., & Yu, L. (2016). Binocular responses for no-reference 3D image quality assessment. *IEEE Transactions on Multimedia, 18*(6), 1077–1084.

Zhou, W., Yu, L., Qiu, W., Wang, Z., & Wu, M. (2016). Utilizing binocular vision to facilitate completely blind 3D image quality measurement. *Signal Processing, 129*, 130–136.

Zhou, W., Qiu, W., & Wu, M. (2017a). Utilizing dictionary learning and machine learning for blind quality assessment of 3-D images. *IEEE Transactions on Broadcasting, 63*(1), 404–415.

Zhou, W., Zhang, S., Pan, T., Yu, L., Qiu, W., Zhou, Y., et al. (2017b). Blind 3D image quality assessment based on self-similarity of binocular features. *Neurocomputing, 224*, 128–134.

Chapter 8
Medical Image Quality Assessment

8.1 Introduction

Medical image quality assessment plays an important role in not only the design and manufacturing processes of image acquisition and processing systems, but also is critical for comparing and optimizing such as X-ray tube voltage and tube current (Wagner et al. 2007; Krupinski and Jiang 2008; Deng et al. 2015). On the other hand, with the development of medical imaging technologies, the assessment on the impact and benefit on patient care is rising (Cavaro-Ménard et al. 2010). Development and design of those medical imaging technologies should take the concept of image quality into account.

In last decade, many successful natural image quality assessment (IQA) tools, such as SSIM, PSNR have been introduced into medical image quality evaluation. Unfortunately, unlike classical distortion types of natural images, medical images are totally different, which leads to low accuracy of simply applying off-the-shelf natural IQA methods. For example, in medical imaging, image quality is governed by a variety of factors such as contrast, resolution (sharpness), noise, artifacts, and distortion. And the quality of CT image is determined by imaging methods, X-ray dose, the characteristics of CT equipment, and the imaging variables selected by operators. Consequently, studying on dedicated medical image quality assessment is strongly required and it can be also utilized as a guide to evaluate and improve the performance of medical imaging and processing algorithms (Liu et al. 2012; Hua et al. 2015). Moreover, for aiding clinicians in rendering a diagnosis, it is widely accepted that image quality should be as good as possible to optimize diagnostic decisions (Zhang et al. 2012, 2014). In medical diagnosis, any visual artifact in the reconstructed CT images may hinder diagnostic conclusions and lead to severe unpredictable consequences. Preserving diagnostic relevant information and maintaining acceptable image quality turn out to be critical in CT imaging system, but there are a lot of problems to be settled in evaluating CT image quality. First, CT image quality is not a single factor but a composite of the five factors

© Zhejiang University Press, Hangzhou and Springer-Verlag GmbH Germany 2018   215
Y. Ding, *Visual Quality Assessment for Natural and Medical Image*,
https://doi.org/10.1007/978-3-662-56497-4_8

mentioned above (Sutha and Latha 2011). As a result, the measurement must be applied in global conditions. Second, huge volumes of CT image data keep clinicians away from subjective assessment. Nowadays, computer assistant models have been developed for the medical image (including CT image) quality assessment to avoid costly clinician observer assessments (Leng et al. 2013; Goossens et al. 2012; Zhang et al. 2015).

Toward these goals, medical image assessment should consider several factors involved in image analysis and interpretation, since it is a multifaceted process that can be approached from a variety of perspectives. However, the researchers have to keep in mind that clinicians need not the most "beautiful" image, but rather the image that offers the most useful and effective presentation for physicians to improve their detection and classification of disease. Studies of medical image assessment should consider not only the image perception but also the critical elements of improving health care by interaction with clinicians' examinations (Cavaro-Ménard et al. 2010).

At present, the development of the medical image assessment is unmatured. Similar to the natural image assessment, it can be categorized as subjective quality assessment and objective quality assessment. Subjective quality assessment methods are run by a group of experts who rate the quality of the diagnosis as well as the visual quality of the image. Cavaro-Menard et al. review the main methodologies to evaluate diagnostic quality of medical images by subjective assessment including ROC analysis and diagnostic criteria quality analysis (Cavaro-Ménard et al. 2010). For objective image quality assessment, a number of methods have been proposed. Various quality measures such as mean square error (MSE), signal-to-noise ratio (SNR), peak signal-to-noise ratio (PSNR), contrast-to-noise ratio (CNR), and contrast improvement ratio (CIR) are commonly used for quality assessment. The modulation transfer function (MTF) (Samei et al. 2006) and noise power spectrum (NPS) (Dobbin et al. 2006) are used to describe the resolution and noise properties of imaging systems, respectively (Deng et al. 2015). And the detective quantum efficiency (DQE) (Neitzel et al. 2004) is used as a metric to represent the general quality of the imaging system. Specially, Tsai et al. propose an image quality metric based on mutual information (MI) (Tsai et al. 2008) to assess the overall image quality of medical imaging systems. Since compression is increasingly used in medical applications to enable efficient and universally accessible electronic health records, Cosman et al. evaluate the quality of compressed medical image with three approaches: SNR, subjective rating, and diagnostic accuracy (Cosman et al. 1994). Khieovongphachanh et al. evaluate the image quality by PSNR and standard deviation (Khieovongphachanh et al. 2008). Recently, Pambrun et al. investigate the relation between CT acquisition parameters and image fidelity with JPEG2000 compression (Pambrun and Noumeir 2013). In fact, the choice of assessment methods depends on the application task that is to be evaluated. In other words, it depends on the information that is extracted from the images themselves (Cavaro-Ménard et al. 2010).

This book has no intention to give a detailed overview on the methods of medical image quality assessment. Instead, this book focuses on: (1) presenting a quality assessment method for portable fundus camera photographs in Sect. 8.2; (2) putting forward a generalized relative quality assessment scheme which is derived from Bayesian inference theory in Sect. 8.3; (3) proposing an adaptive paralleled sinogram noise reduction method for low-dose X-ray CT in Sect. 8.4, as a case study and application of the relative quality assessment; (4) studying on the relationship between the image quality and imaging dose in low-dose CBCT based on dose-quality maps in Sect. 8.5.

8.2   IQA of Portable Fundus Camera Photographs

The first case of applying IQA in biomedical imaging concerns evaluating the quality of portable fundus camera photographs. Since we are in the era of tele-medicine in blossom and the medical "big data" in aggrandizing, technical advances have spread over numerous domains. Specifically, ophthalmology is now highlighting the use of nonmydriatic ocular fundus photography, which gives rise to applications of portable fundus cameras (Jin et al. 2017; Kawaguchi et al. 2017). Figure 8.1 illustrates a sample of a portable fundus camera (DEC200 manufactured by the Med-imaging Integrated Solution Inc.) and several fundus images. Compared with traditional fundus photography machines that are on a permanent fixture, a portable digital fundus camera is held on operators' hands and needs no mydriatic participation. Nonmydriatic means that there is no requirement of pupil dilation. Therefore, it seems to be a promising solution in serving for districts or remote areas lacking medical resources. However, such hands-on operating

(a)   (b)   (c)

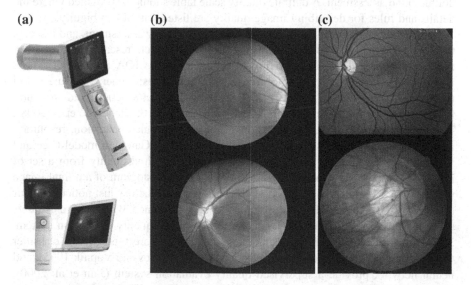

Fig. 8.1 Illustration of a portable fundus camera (**a**), two images from proprietary datasets (**b**), and two images from public datasets (**c**)

conditions are destined to be vulnerable to image quality degradation. Typical distortion comes from uneven luminance distribution, fluctuations in focus, and patients' movements, etc., which prohibits routine detection and diagnosis on vision impairment. Hence, grading the image quality of portable fundus camera imaging system is of great value.

8.2.1 Basic Problem and Model

Since we focus on evaluating medical image quality, there are two basic problems:
1. How can we get subjective assessment on specific groups of medical images?
2. How can we develop an automatic system to give relevant objective scores that fit with subjective ones?

The first problem is actually a piece of cake in natural IQA for natural image quality databases are always available, but in medical image IQA domain, the situation is different.

Images from particular groups of patients should be regarded as propriety databases that cannot be shared on the Internet. The variety of modalities as well as contents relevant to different body parts add difficulties for subjective assessment. As for portable fundus photography, no databases have been released so far, and in this case studying, we used a proprietary database that was connected by two DEC200 portable fundus cameras. Next, we can call upon ophthalmologists to classify and grade these images. To be honest, this is a systematical engineering. You need at least two ophthalmologists to do the job. They should work independently, named double-blind assessment. A definite quality scale table should be provided where the details and rules for describing image quality are listed with no ambiguity.

The second problem is obvious in natural image quality assessment, and it seems that we have a series of algorithms to solve it. In fact, researches have been conducted over last decades, leading to different fundus IQA algorithms. These algorithms can be grouped into two major categories: generic feature-based methods (Marrugoa et al. 2015; Dias et al. 2014; Fasih et al. 2014) and structural feature-based methods (Şevik et al. 2014; Paulus et al. 2010; Giancardo et al. 2008). The former deals with global distortions, such as luminance distortion, resolution artifact, or limited contrast. Explicit templates such as a Gaussian model (Lee and Wang 1999) are first proposed to extract images of desired quality from a set of images. Convolution of the template with the intensity histogram of a retinal image was taken as generic quality. Later, HVS characteristics such as just noticeable blur (Narvekar and Karam, 2011) are employed and combined with texture features (Yang et al. 2002) to constitute a generic retinal image quality estimation system. Meanwhile, machine learning techniques such as k-nearest neighbors classifier (Keller et al. 1985), support vector machine (SVM) (Cortes and Vapnik 1995), and neural network provide a supervised quality evaluation system (Jain et al. 2000). The latter pay attention to particular distortions on specific locations on fundus images. Usually, these need to segment optic disk, macular, or vessels, followed by extracting quality-aware features from these regions and mapping these features into objective quality.

In the following, the two basic problems will be solved by combining medical purposes and human vision system.

8.2.2 Subjective Quality Assessment

Three ophthalmologists, one senior (marked by S) and two juniors (marked by J1, J2), were called upon to operate subjective assessment on 536 portable fundus photographs. They screened a 0.275 mm per pixel monitor with the viewing distance of about 30 cm. Meanwhile, a generic quality gradation scale is listed in Table 8.1.

This scale tells us that image quality of portable fundus photography consists of three elements: luminance, sharpness, and contrast. It is different from other schemes like Fleming's image clarity grading scheme (Fleming et al. 2006), which concerned small vessel clarity, the field definition of optic disk and macular.

The three elements form a three-bit binary number so that a grading instance can be enumerated as the following: 000, 001, 010, 100, 101, 110, 111. For each element/digit, the unit "1" indicates that the principle description is agreed with the image whereas the null (0) denotes the opposite situation. Each image is associated with the three-digit binary subjective score which indicates whether the image suffers from the uneven luminance or color (Lumc, told by the left bit), sharpness distortion (SD, told by the middle bit), and low contrast distortion (LC, told by the right bit).

As there were three ophthalmologists grading the subjective generic quality, the consistency of inter-ophthalmologist grading should be investigated. The Spearman rank-order correlation coefficient (SROCC) was chosen as the consistency indicator which is widely used in statistical analysis. For each grading bit in Table 8.2, the SROCC between S and J1, S and J2, J1 and J2 were calculated. As there are three bits, a 3-by-3 SROCC consistency matrix is shown in Table 8.2.

Table 8.1 Generic quality gradation scale

Grading bits	Principle description
0/1 _ _	Image is with the uneven illumination or color
_ 0/1 _	Image has noticeable blur appeared in optic disks, vessels, or background, etc.
_ _ 0/1	Image's contrast is low, so the band of pixel intensity is narrow

Table 8.2 Consistency matrix of interobserver grading

SROCC between	S and J1	S and J2	J1 and J2
Grading bit 1 (0/1 _ _)	0.6531	0.5853	0.6812
Grading bit 2 (_0/1_)	0.6967	0.8016	0.7074
Grading bit 3 (_ _ 0/1)	0.5517	0.5805	0.4703

8.2.3   Objective Quality Assessment

After subjectively grading all images, we obtained a redefined database which consists of images with associated labels. In detail, there are totally 302 images; as for Lumc, there are 66 images with acceptable quality and 236 images with degraded quality. For convenience, images with acceptable quality are named positive samples while images with low quality are negative samples. In this way, the distributions of the three partial quality targets, as well as the overall quality targets, on the database are shown in Table 8.3. The numbers refer to positive and negative volumes.

As an illustration, Fig. 8.2 displays four groups of portable fundus images in the database. They are arranged in four rows, representing negatives from Lumc, SD, LC, and positives from overall.

8.2.4   Feature-Based Objective IQA Framework

Since there are three partial distortion types and one overall distortion type, all of which respond to a two-classes classification task, feature-based and machine learning founded methodology is adopted. Figure 8.3 outlines the major methodological steps.

It can be seen that the methodology consists of three steps: The preprocessing step aims to remove irrelevant background to save computation costs; the second step focuses on the feature extraction that is composed of three sub-models: multichannel sensation, just noticeable blur, and contrast sensitivity function; the last machine learning step was devoted to evaluating the algorithm's capability of binary classification of images. Two machine learning tools, the SVM and the decision tree (DT), were employed. We will introduce these steps in detail in the following sections.

8.2.5   Preprocessing

On the left and right side of our fundus images, there are large area of black regions which did no benefit for next processing. What's more, the resolution of a fundus image is 2560×1960 pixels. If one pixel is recorded by three bytes (each byte recording each color channel), it needs about 14 MB for loading an image. As a consequence, 100 images will occupy nearly 1.5 GB. Remember that it is just the

Table 8.3 Distribution of subjective targets on the collected database		Lumc	SD	LC	Overall
	Positive	66	149	176	39
	Negative	126	153	126	263

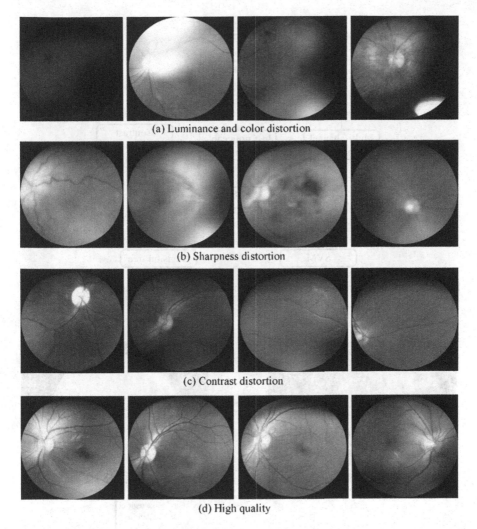

(a) Luminance and color distortion

(b) Sharpness distortion

(c) Contrast distortion

(d) High quality

Fig. 8.2 Illustration of portable fundus photographs of four groups

beginning; IQA algorithms must be allocated with extra memory to keep temporary variables. This high complexity of space obliges us to employ preprocessing strategies. Firstly, notice that content of a fundus image is in a circle-like region in the center; hereafter, the well-known thresholding method called OTSU is employed. Luminance intensities larger than the OTSU threshold are taken as the foreground area. Generally, the area is connected so that centroid and associated radius can be calculated. As a result, the circle-like profile is plotted and cropped. With a following down-sampling step, the size of a fundus image was rescaled to 512×512 so that the memory for loading is only 0.75 MB. Figure 8.4 summarizes these preprocessing steps.

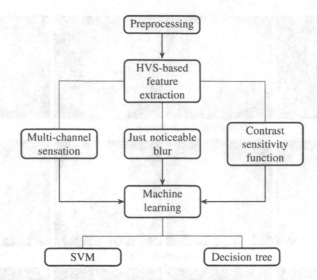

Fig. 8.3 Framework for objectively evaluating portable fundus photographs

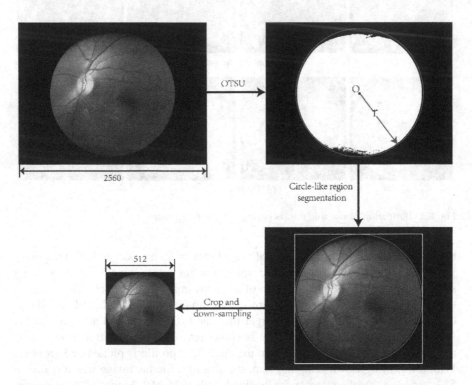

Fig. 8.4 Preprocessing on portable fundus images

8.2.6 Feature Extraction

In previous chapters concerning natural image quality assessment, we have reviewed and discussed human visual system (HVS) which makes objective quality to be perceptual. Here, we apply three characteristics in the fundus IQA. Why three? It is not a chance but a motivation that each HVS characteristic builds a model for one partial distortion type, so we take advantage of multichannel sensation for Lumc, just noticeable blur for SD, and contrast sensitivity function for LC.

8.2.6.1 Multichannel Sensation

Studying on HVS tells us that human perceives luminance and color in different channels. Rods and cones in retina are sensitive to brightness and darkness, respectively. Three subtypes of cones are responsible for light rays in long, middle, and short frequency intervals accordingly, which enable perception of colors. In our circumstance, luminance suffers from overexposure and underexposure; color distortion mainly refers to queer hue. To detect the three aspects, we transformed the RGB (R: red, G: green, B: blur) channels into HIS (H: hue, I: intensity, S: saturation) channels using a color space transformation matrix (Toet and Lucassen 2003):

$$
\begin{pmatrix} I \\ H \\ S \end{pmatrix} = \begin{bmatrix} 0.299 & 0.587 & 0.114 \\ 0.596 & -0.274 & -0.322 \\ 0211 & -0.523 & 0.312 \end{bmatrix} \begin{pmatrix} R \\ G \\ B \end{pmatrix} \tag{8.1}
$$

where the pixel set of the intensity component is denoted as $I(x, y)$, $1 \leq x \leq M$, $1 \leq y \leq N$. The rows and cols of an image is M and N, respectively. Similarly, the previously preprocessed image mask can be expressed as $MI(x, y)$, $1 \leq x \leq M$, $1 \leq y \leq N$. Note that $MI(x, y)$ is a 0/1 binary image. To quantify underexposure and overexposure, double threshold masking is employed on the intensity channel. The lower threshold T_{lower} is given by

$$
T_{lower} = \frac{1}{n} \sum_{i=1}^{n} \{ I(x_i, y_i) | MI(x_i, y_i) = 0 \} \tag{8.2}
$$

where n is total number of the pixels at which $MI(x, y) = 0$. The upper threshold T_{upper} is related to T_{lower} and calculated as

$$
T_{upper} = 255 - \frac{T_{lower}}{4} \tag{8.3}
$$

Now the illumination mask $MI_1(x, y)$ can be obtained as

$$MI_1(x, y) = \begin{cases} 1, & T_{\text{lower}} \leq I(x, y) \leq T_{\text{upper}}, \\ 0, & \text{else.} \end{cases} \tag{8.4}$$

Similarly, to perceive queer color, a relevant mask $MI_2(x, y)$ is obtained using single thresholding on the hue component H (see the color space transformation equation before). The threshold T_{color} is defined as

$$T_{\text{color}} = \frac{1}{n} \sum_{i=1}^{n} \{H(x_i, y_i) | MI(x_i, y_i) = 0\} \tag{8.5}$$

where the $H(x, y)$ denotes hue value at (x, y), and the binary color distortion mask is further computed as

$$MI_2(x, y) = \begin{cases} 1, & H(x, y) \geq T_{\text{color}}, \\ 0, & \text{else.} \end{cases} \tag{8.6}$$

Based on applying logical AND operation on MI_1 and MI_2, a mask that takes both luminance distortion and color distortion into consideration is denoted as MI_{ROI},

$$MI_{\text{ROI}}(x, y) = \begin{cases} 1, & MI_1(x, y) = 1 \& MI_2(x, y) = 1 \\ 0, & \text{else.} \end{cases} \tag{8.7}$$

By counting the logical true pixels in MI_{ROI} and computing the ratio of the number to the size of MI, we can extract the Lumc-aware feature,

$$f_1 = \frac{\sum_{x,y=1}^{M,N} MI_{\text{ROI}}(x, y)}{\sum_{x,y=1}^{M,N} MI(x, y)} \tag{8.8}$$

Figure 8.5 gives three image samples as well as the luminance mask, color mask, and fused mask. It can be seen that the luminance mask and color mask compensate for the other, and the fused mask detects both luminance and color distortions.

8.2.6.2   Just Noticeable Blur

JNB is one of the most important and founded theories in human perception (Ferzli and Karam 2009). As for application, it has been employed in natural image quality assessment (Narvekar and Karam 2011) and retinal image sharpness (Fasih et al. 2014). These two pieces of research follow a similar scheme: First, use Sobel edge detection to find an edge, then calculate the edge width. Since spatial edge detection on images is equivalent to apply mathematical differentiation on an image, which

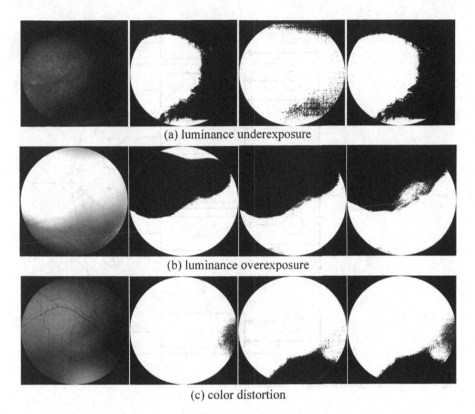

(a) luminance underexposure

(b) luminance overexposure

(c) color distortion

Fig. 8.5 Multichannel distortion examples: Each row represents one particular distortion type; from left to right: the RGB image, luminance mask, color mask, and the fused mask, respectively

avoidably introduces noise artifacts. What's more, edge width calculation is time consuming because the widths of edge segments vary. To handle this drawback of blur assessment in our portable fundus photography, we applied JNB on edge density, which assumes that blur distortion has apparent visual effects on vessels of retina. Our pipeline of calculating JNB on vessel density is:

1. Extraction of vessels on the fundus images;
2. Transforming vessel densities in gray scale to probabilistic scale map;
3. Calculation of JNB on a probabilistic histogram of the vessel density map.

To extract vessels or their edges on the fundus images, a top-hat algorithm is utilized where twelve orientations of linearly dilation templates are employed (Zana and Klein 2001). Angle increases from $0°$ to $165°$ with the step of $15°$. Figure 8.6 outlines a scheme of the top-hat algorithm.

The input image denoted by I in Fig. 8.6 is a grayscale image, and the output map denoted by E is a vessel density map whose intensities range from 0 to thousands as a result of summation in the last step of the scheme. So far, the first step of the pipeline is finished. For the second step, we need to scale down the map

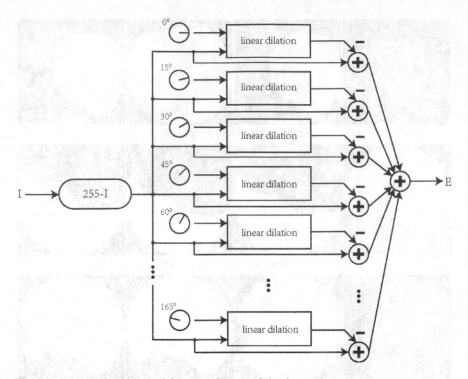

Fig. 8.6 Top-hat algorithm used for extracting vessel density maps

into a probabilistic one where all intensities are within 0 and 1. Similar to cumulative probability blur detection [CPBD (Narvekar and Karam 2010)], a formula to achieve this is as follows:

$$P(x, y) = 1 - \exp\left(-\left|\frac{E(x, y)}{E_{\text{JNB}}}\right|\right) \qquad (8.9)$$

where $E(x, y)$ and $P(x, y)$ refer to original density map and probabilistic map, respectively. The prior variable E_{JNB} has an effect on restraining noise. Empirically, it is determined by the standard deviation (std) of all the original densities:

$$E_{\text{JNB}} = \begin{cases} 110, & \text{std}(E) \leq 13, \\ 120, & \text{else.} \end{cases} \qquad (8.10)$$

Now the second step is finished, and as an illustration, we display three fundus images with different degrees of blur distortion and corresponding probabilistic maps in Fig. 8.7. It can be seen that as blur distortion degree increases, the edge densities in probabilistic maps decrease accordingly.

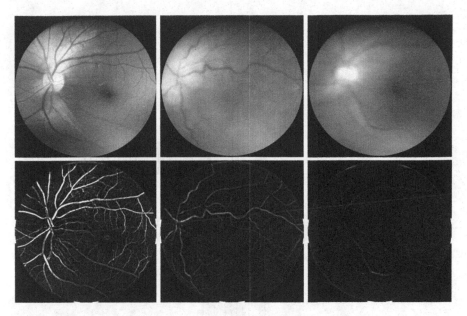

Fig. 8.7 Blur distortion examples: retinal images with increasing blur distortion degree at the top row and vessel density probabilistic maps at the bottom row

To complete the JNB pipeline, a histogram profiling distribution of the probabilistic map is plotted. Based on the histogram, JNB index is defined by regions on the right side of a horizontal threshold P_{JNB}. The threshold can be easily determined by making $E(x, y)$ equals to E_{JNB} in Eq. (8.9); therefore P_{JNB} approximately equals to 0.632. Figure 8.8 illustrates three histograms responding to Fig. 8.7.

The JNB index can be mathematically expressed by

$$\text{JNB} = p(P(x,y) \geq P_{JNB}) = \sum_{P_{JNB}}^{1} p(P(x,y)) \tag{8.11}$$

where p denotes the frequencies of histograms scaled within 0 and 1. It can be observed that as the degree of blur distortion increases, the green regions shrinks, indicating the JNB index is competent to detect blur distortion. Hence, we denote JNB as the blur-aware feature f_2.

8.2.6.3   Contrast Sensitivity Function

The last partial distortion is low contrast. Traditional contrast evaluation methods aim to estimate a standard deviation of image intensities directly, which ignores contrast perception mechanism, i.e., contrast sensitivity function (CSF). Here, we introduced an improved contrast evaluation method where CSF is employed. In

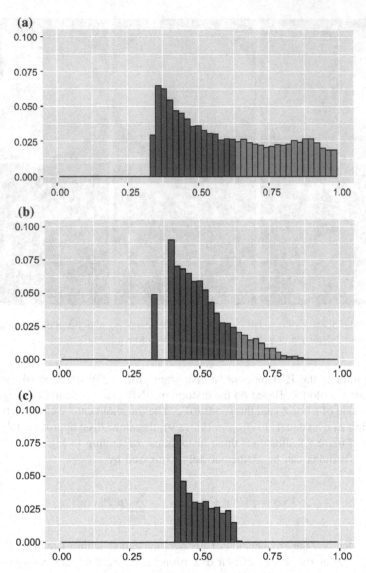

Fig. 8.8 Histograms of vessel density probabilistic maps where three JNB indices corresponding to increased blur distortion degree are denoted by green regions

fact, a typical workflow of using CSF to predict image quality has been built over decades (Ginesu et al. 2006; Yu and Cai 2014), and numerous CSF formulas have been summarized (Yu and Cai 2014). Considering simplicity, we chose the formula as follows:

$$\text{CSF}(r; a_0, a_1, a_2) = \left(1 - a_1 + \frac{r}{a_0}\right) \exp\left[-\left(\frac{r}{a_0}\right)^{a_2}\right] \qquad (8.12)$$

where a_0, a_1, and a_2 are parameters that should be set explicitly, and r is spatial frequency measured in circle per degree (cpd). Note that this is a CSF model for one-dimensional signal. A two-dimensional image has orthogonal frequency components denoted by u and v. All the three variables are ruled by the Euclidean theorem:

$$r = \sqrt{u^2 + v^2} \qquad (8.13)$$

where r is always positive, and it is also obvious that one-dimensional CSF curve looks like a bandpass filter. The center frequency, as confirmed by (Ginesu et al. 2006), is nearly 11.3 cpd, at which the response of CSF approaches the maximum. To meet this requirement, a_0, a_1, and a_2 are set to 11.3, 1, and 1, respectively. Then comes the question: how to utilize the well-configured CSF in contrast evaluation? Like JNB-based blur evaluation, there is also a pipeline:

1. Use the fast Fourier transform (FFT) to transform an image into spatial frequency domain;
2. Use the CSF to weight the spatial frequency coefficients;
3. Use the inverse fast Fourier transform (IFFT) to transform the CSF-weighted spatial frequency coefficients back into spatial domain.

This pipeline can be expressed mathematically as an equation:

$$\widehat{I}(x, y) = \text{IFFT}\{\text{FFT}\{I(x, y)\} \times \text{CSF}(r)\} \qquad (8.14)$$

where the symbol $\times$ means that coefficients of FFT are multiplied by corresponding CSF response at circle-like grids since spatial frequency defined in Eq. (8.13) is actually a radius in polar coordinates. The $I(x, y)$ is the grayscale image, and the $\widehat{I}(x, y)$ is the output image of IFFT. Figure 8.9 illustrates the pipeline.

You may wonder that whether or not the spectrum of an actual fundus image can cover center frequency (11.3 cpd). Don't worry, you can check it by calculation! First, note that the tangent of 1 deg equals to 1.7455; therefore, 1 cpd can be explained that an approximate 1.75 cm wide pattern subtended an angle of 1 degree when we viewed the pattern at the distance of 1 m. Second, digital and physic resolutions of a portable fundus image are 2560×1960 and 0.275 mm/pixel, respectively. Third, the viewing distance from the observer to the monitor is 30 cm. Pooling all conditions together, the maximum spatial frequency (r_{max}) can be calculated as

$$r_{max} = \frac{\sqrt{2560^2 + 1960^2} \cdot 0.275}{1.7455 \cdot 10 \cdot 30 \div 100} = 169.3\,\text{cpd} \qquad (8.15)$$

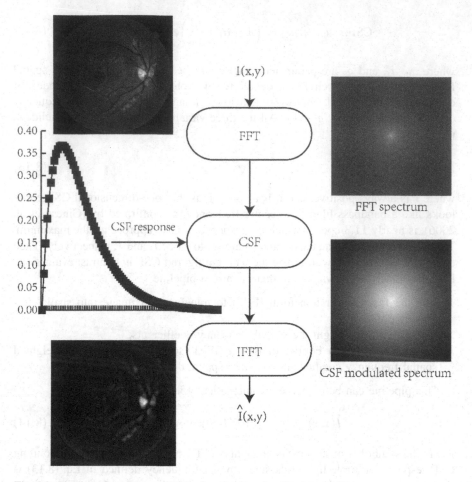

Fig. 8.9 Pipeline of contrast sensitivity function-based contrast evaluation

Without doubt, it is much larger than the center frequency. In practice, the monitor used for display may not have the enough resolution as fundus images do, so r_{max} is also limited by a monitor's size so that the maximum frequency is the minimum of results based on the two circumstances.

Finally, the output image $\widehat{I}(x, y)$ was then normalized within 0 and 1, and the contrast sensitivity feature f_3 was computed by the following equation:

$$f_3 = \frac{Q_3\left(\widehat{I}\right) - Q_1\left(\widehat{I}\right)}{\max\left(\widehat{I}\right) - s\left(\widehat{I}\right)} \tag{8.16}$$

where Q_1 and Q_3 are the first and third quartile values of a boxplot of $\widehat{I}(x, y)$, and the denominator referred to the range of $\widehat{I}(x, y)$ with outlier values excluded. As an

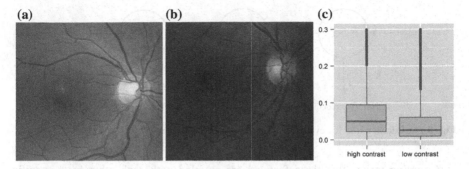

Fig. 8.10 Contrast evaluation based on CSF: **a** retinal image with high contrast, **b** retinal image with low contrast, **c** boxplot of the CSF-processed coefficients

illustration, Fig. 8.10 presents two images with high contrast, low contrast, and their boxplot.

8.2.7 Machine Learning

Since we have obtained three features, f_1, f_2, and f_3, to quantify the quality of Lumc, SD, and CD, respectively, the last thing to do is mapping them into an objective quality. In natural image quality assessment, it is easy for us to fit a one-variable nonlinear regression function because there is only one quality-aware feature. In medical image quality assessment, that is not the same thing. We need to evaluate medical image quality in an omni-perspective way such that a multiple variables function must be optimized. Without loss of generality, we can define the function applied in our fundus IQA as

$$y = F(f_1, f_2, f_3; P) \tag{8.17}$$

where P is a set of inner parameters of the function F, and y is the output that is actually a three-elements vector corresponding to subjective grading format. For structure of the function F, there are two basic forms: One is resoluble and the other is coupled. Resolvability means that F can be decomposed into several one-variable sub-functions:

$$F(f_1, f_2, f_3; P) = \sum_{i=1}^{3} F_i(f_i, P_i) \tag{8.18}$$

which can be implemented by a decision tree (DT) that has three nodes, each predicts one partial distortion. Figure 8.11 plots the decision tree for our fundus IQA.

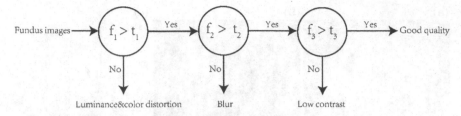

Fig. 8.11 Decision tree for evaluating partial and overall distortion

Coupling means that three features are coupled with each other by a specific operation. SVM is a member of this kind, using kernels like radial basis function to couple features. Kernel functions, such as a polynomial function or a radial-based function (RBF), are often used to transform the original feature space into a higher dimensional space. For convenient, lots of excellent SVM toolboxes like LIBSVM (Chang and Lin 2011) and Kernlab package (Zeileis et al. 2004) have been released where relevant kernel parameters are automatically configured.

Both DT and SVM have to be trained before predicting objective image quality. To train DT, we should optimize three thresholds (t_1, t_2, t_3) so that the sensitivity and specificity are the highest. The sensitivity and specificity are defined as below,

$$\text{sensitivity} = \frac{TP}{TP + FN} \tag{8.19}$$

$$\text{specificity} = \frac{TN}{TN + FP}. \tag{8.20}$$

where TP denotes true positive samples, and TN denotes true negative samples; FP and FN denote false positive and false negative, respectively. This optimization problem can be converted into a minimum distance searching problem on receiver operator characteristics (ROC) curves. To do this, a n-fold cross-validation is employed, where $(n - 1)$ folds of dataset are used for optimizing the three thresholds and the remaining fold for testing. Note that different thresholds aim to predict different partial distortion, so do the subjective labels. Figure 8.12 visualizes

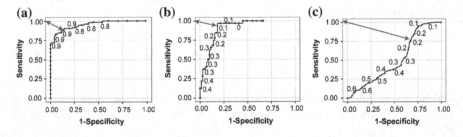

Fig. 8.12 Decision tree parameters optimization process for Lumc, SD, and CD, respectively

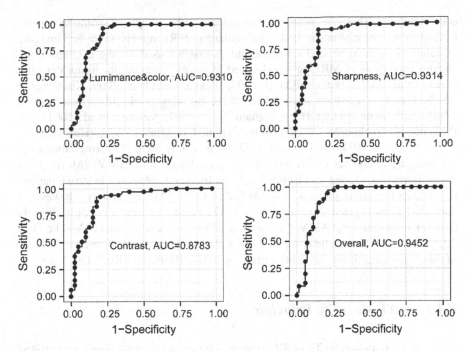

Fig. 8.13 Area under curve (AUC) of SVM for partial and overall distortion types

the searching process for the three thresholds where the minimum distance is found on the nearest point on a ROC curve to the top-left vertex.

Training a SVM is actually a supervised binary classification problem. Similar to DT, a n-fold cross-validation is employed. But here we are not going to search an optimized threshold. Specifically, SVM defines a cost function that should be minimized for two-classes prediction. Figure 8.13 presents the cross-validation-based prediction performance of SVM for three types of partial distortions and the overall distortion.

8.3 IQA of Susceptibility Weighted Imaging

It is well known that medical imaging concerns more about invisible light. As visible light fails to pass through various human tissues like bones, vessels, or fat, physicians and engineers employ rays of higher frequency such as X-ray (Chapman et al. 1997) and ultrasound (Szabo 2004). Alternatively, a more advanced type of imaging technique takes advantage of an abundance of hydrogen in our body tissues, which is called magnetic resonance imaging [MRI (Acharya and Ray 2005)]. First conceived by noted physicists Bloch and Purcell in the year 1946, MRI has derived various techniques such as spin echo imaging, gradient recalled echo imaging. They all prone to the principle that magnetic stimulus excites hydrogen nuclei in human body and emanate MRI signals used for imaging. Differences of them depend upon the pulses that are used for the excitation, and various modalities of MRI are widely used in visceral imaging for tumor detection (Lichy et al. 2007; Beyersdorff et al. 2005; Iftekharuddin et al. 2009; Kircher et al. 2012), blood oxygenation (Tian et al. 2010; Bohning et al. 2001; Ogawa et al. 1990), soft tissue injuries (Cunningham et al. 2007; Vaccaro et al. 2001), etc. Particularly, single-echo susceptibility weighted imaging (SWI$_s$) (Haacke et al. 2009; Yan et al. 2012a, b; Horie et al. 2011) is mainly applied in stroke diagnosis. An imaging biomarker called asymmetric medullary veins (AMV) (Morita et al. 2008; Jensen-Kondering and Böhm 2013) are frequently observed in stroke patients rather than in normal patients, and SWI$_s$ can detect AMV accurately. However, it needs to be investigated which time of echo (TE) on single-echo susceptibility is the optimal for visualizing AMV (Wang et al. 2016; Horie et al. 2011; Mucke et al. 2015). Also, it needs to be compared the ability to detect AMV between SWI$_s$ and multi-echo susceptibility weighted imaging [SWI$_c$ (Denk and Rauscher 2010)].

8.3.1   SWI Modality Principles

Different from traditional T1 or T2 imaging, SWI deploys a fully flow compensated, long TE, gradient recalled echo (GRE) pulse sequence to acquire a type of contrast image. Apart from exploiting phase image to detect the susceptibility differences between tissues, SWI used magnitude data to produce an enhanced contrast magnitude image. For SWI$_s$, it has been reported that optimal venous contrast of a vein is achieved at TE = 28 ms (Reichenbach et al. 2000; Koopmans et al. 2008). Similarly, for investigating the best image contrast of AMV, three echoes including TE of 23.144, 29.192, and 35.240 ms are used to reconstruct SWI$_s$. On the other hand, SWI$_c$ images are proposed to have better contrast-to-noise ratio and signal-to-noise ratio than SWI$_s$ images (Denk and Rauscher 2010), so what we are going to do is first acquiring three SWI$_s$ images followed by averaging the three different echoes of SWI$_s$ to get SWI$_c$. The processing flow of three SWI$_s$ and one SWI$_c$ is sketched in Fig. 8.14. Note that the SWI$_s$ image is created by combining the magnitude and high-pass (HP) filtered phase images. The phase mask is created from a normalized phase image by mapping all values above 0 radians to be 1; for values below 0 radians, a polynomial mapping function is called to regularize them to be within 0 and 1. The magnitude image is then multiplied point-to-point by this mask.

8.3.2   Subjective Quality Assessment

Before our objective quality assessment on single- and multiple-echo SWI images, the subjective quality standard should be provided. Here, we collected the four

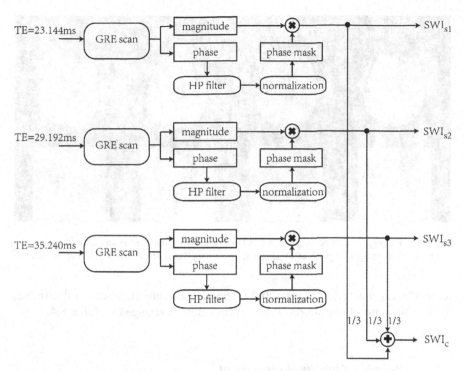

Fig. 8.14 Reconstruction flow of three single-echo time (TE) SWI and one multi-echo SWI. As an illustration, Four SWI slices from a patient are displayed in Fig. 8.15; the region of AMV is zoomed in for better visualization

series of SWI images from 14 stroke patients. The recruitment criteria are as follows: (1) to be first-ever ischemic stroke; (2) with the vascular territory of unilateral middle cerebral artery involved; (3) admission about 3–7 days after stroke onset; (4) above 18 years old; (5) without hemorrhagic infarction; (6) no history of neurological disorders. Each patient is scanned using 3D high-resolution flow-compensated multi-echo SWI sequence on a 3.0T MR scanner (TE = 23.144–35.240 ms with an echo spacing of 6.048 ms, flip angle = 25°, bandwidth = 41.67 kHz, slice thickness = 2 mm, matrix = 384 × 320, field of view = 24 cm). To further increase in-plane resolution, 384 × 320 images are interpolated cubically into 512 × 512 images. As a result, here we collect 14 groups of SWI images, with four types of modality in each group and with 8 plane slices in each modality. Further, the sixth slices of the 3D SWI images are extracted for quality assessment, so there are 56 (14 × 4) slices to be evaluated.

Next, two radiologists who are evaluated blinded to the other are invited to give an opinion score from a five-point scale to each image. The five-point scale concerns the definition for identifying AMV. It explains as follows: A score of 1 indicates non-diagnostic; a score of 2, poor; a score of 3, moderate; a score of 4, good; and a

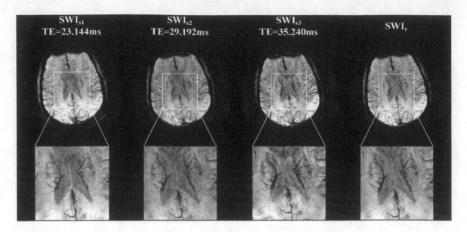

Fig. 8.15 Examples of three single-echo and one combined multi-echo susceptibility weighted imaging (SWI) images, regions of asymmetric medullary veins are zoomed in at bottom row

score of 5, excellent (Ghrare et al. 2008). The standard subjective score is determined by the minimum of two scores. Final scoring table is arranged in Table 8.4.

8.3.3   Relative Quality Assessment

Why relative? In previous sections, we have been familiar with full-reference (FR) IQA, reduced-reference (RR) IQA, and no-reference (NR) IQA. The first two need both original image/features and distorted image/features, and the last one needs prior knowledge about distortion types. The first example in our medical quality assessment, portable fundus photography quality evaluation, achieves high performance because of the well-built prior knowledge-based model. In this application, there are no original SWI images which are invulnerable to distortion, and the prior knowledge of specific distortion of AMV is yet unavailable. Therefore, FR, RR, and NR are not applicable. Fortunately, we have at least one piece of prior knowledge: If images are different, their quality must be different and sortable. This brings the concept of relative quality which highlights not the absolute objective scores but the rank order of the scores of images to be compared (Wang et al. 2014).

Table 8.4 Subjective scores of susceptibility weighted imaging images

	01	02	03	04	05	06	07	08	09	10	11	12	13	14
SWI_{s1}	4	3	3	3	3	3	3	4	3	4	3	3	3	3
SWI_{s2}	4	4	4	4	3	4	4	4	4	4	4	4	4	4
SWI_{s3}	4	3	3	4	3	3	3	4	3	4	3	3	3	3
SWI_c	5	5	5	5	5	5	5	5	5	5	5	5	5	5

Without loss of generality, suppose we have N images to calculate their relative quality, we can model the constant of the sum of their quality according to probability theory:

$$P\left(\prod_{i=1}^{N} I_i\right) = \text{Const} \tag{8.21}$$

where I_i denotes the ith image. To calculate quality of an image, say I_i, relative to other images, we can apply conditional probability equation as following,

$$P\left(\prod_{i=1}^{N} I_i\right) = P(I_1 I_2 \cdots I_{i-1} I_{i+1} \cdots I_N | I_i) P(I_i) \tag{8.22}$$

In the same way, the equation is correct for the first image I_1,

$$P\left(\prod_{i=1}^{N} I_i\right) = P(I_2 \cdots I_{i-1} I_i I_{i+1} \cdots I_N | I_1) P(I_1) \tag{8.23}$$

By default, we calculate the relative quality of I_i to I_1 as

$$P(I_i) = \frac{P(I_2 \cdots I_{i-1} I_i I_{i+1} \cdots I_N | I_1)}{P(I_1 I_2 \cdots I_{i-1} I_{i+1} \cdots I_N | I_i)} P(I_1) \tag{8.24}$$

Since all SWI images are reconstructed independently, the combined probability can be factorized by relevant marginal probability,

$$\frac{P(I_2 \cdots I_{i-1} I_i I_{i+1} \cdots I_N | I_1)}{P(I_1 I_2 \cdots I_{i-1} I_{i+1} \cdots I_N | I_i)} = \frac{P(I_2 | I_1) \cdots P(I_{i-1} | I_1) P(I_i | I_1) P(I_{i+1} | I_1) \cdots P(I_N | I_1)}{P(I_1 | I_i) P(I_2 | I_i) \cdots P(I_{i-1} | I_i) P(I_{i+1} | I_i) \cdots P(I_N | I_i)} \tag{8.25}$$

Note that the probability $P(I_i | I_j)$ represents the quality of I_i corresponding to I_j, which can be calculated by any full-reference algorithms. Considering a probability value should be ranged from 0 to 1, we map full-reference objective quality to the probability by a negative exponential function,

$$P(I_i | I_j) = e^{-f(I_i, I_j)} \tag{8.26}$$

where $f(I_1, I_2)$ denotes the objective quality index output by a FR algorithms. Now Eq. (8.24) can be rewritten as

$$P(I_i) = \frac{e^{-[f(I_2, I_1) + \cdots + f(I_{i-1}, I_1) + f(I_i, I_1) + f(I_{i+1}, I_1) + \cdots + f(I_N, I_1)]}}{e^{-[f(I_1, I_i) + f(I_2, I_i) + \cdots + f(I_{i-1}, I_i) + f(I_{i+1}, I_i) + \cdots + f(I_N, I_i)]}} P(I_1) \tag{8.27}$$

It is obvious that when $i = 1$, the fractal item is simplified to be 1, which means that this item is the key to relative quality. For convenience, we denote this item as $K\,(I_i, I_1)$ which is the relative quality of I_i to I_1.

Return to our SWI image quality assessment. For each patient, we have four SWI images (SWI_{s1}, SWI_{s2}, SWI_{s3}, and SWI_c) to be evaluated, so N equals to 4 in Eq. (8.27). Besides, we choose a classic full-reference algorithm, SSIM, as the internal function. As SSIM (Wang et al. 2004) is symmetric for two input images, i.e., $SSIM(I_i, I_j) = SSIM(I_j, I_i)$, $K\,(I_i, I_1)$ can be further simplified,

$$
\begin{cases}
K(I_2, I_1) = \dfrac{e^{-\left[SSIM(I_3, I_1) + SSIM(I_4, I_1)\right]}}{e^{-\left[SSIM(I_3, I_2) + SSIM(I_4, I_2)\right]}} \\[2mm]
K(I_3, I_1) = \dfrac{e^{-\left[SSIM(I_2, I_1) + SSIM(I_4, I_1)\right]}}{e^{-\left[SSIM(I_2, I_3) + SSIM(I_4, I_3)\right]}} \\[2mm]
K(I_4, I_1) = \dfrac{e^{-\left[SSIM(I_2, I_1) + SSIM(I_3, I_1)\right]}}{e^{-\left[SSIM(I_2, I_4) + SSIM(I_3, I_4)\right]}}
\end{cases} \tag{8.28}
$$

As an illustration, two standard deviation bar histograms compare subjective and relative quality scores across all patients, respectively. It can be seen that relative quality is aligned with the subjective one, indicating the superiority of multiple-echo modality.

8.3.4 Tricks for Relative Quality Algorithm

According to Eq. (8.27), relative quality index is determined by two factors: One is the full-reference kernel function; the other is correlation-like calculation. You must wonder if any full-reference IQA algorithms can be embedded in this equation, or if there is restrictions or conditions on the kernel function. The couple pieces of suspicion give the occasion of tricks for relative quality calculation. There are indeed conditions for the kernel functions. Firstly, value of Eq. (8.26) should be ranged from 0 to 1, which demands that the objective quality index given by a kernel function should be positive. Fortunately, this is somewhat easy to meet since full-reference algorithms calculate either similarity or disparity between two images. However, the mapping of these two criteria to subjective quality is different. Objective similarity is usually positively correlated with subjective quality, while objective disparity is opposite. This without doubt affects the polarity of relative quality. From Fig. 8.16, we can see that relative quality (RQ) from the SSIM-based kernel function is positively related with subjective quality; thus it can be deduced that disparity-based full-reference algorithms such as MSVD (Shnayderman et al. 2006), ASVD (Mansouri et al. 2009), GMSD (Xue et al. 2014) have negative polarity. It would cause ambiguity of the best relative quality if polarity of kernel functions is changed; therefore, the first trick is to regularize the polarity of kernel functions. By default, functions with positive polarity need no change but functions with negative polarity will be regularized by

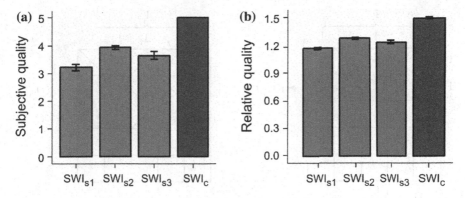

Fig. 8.16 Comparison of standard deviation bar histograms of subjective scores and SSIM-based relative quality indices between three different echoes of single-echo susceptibility weighted images (green histogram) and combined multi-echo SWI images (red histogram). (Color figure online)

$$f(I_i, I_j) = f^{-1}(I_i, I_j) \tag{8.29}$$

Secondly, most kernel functions are symmetrical such that

$$f(I_i, I_j) = f(I_j, I_i) \tag{8.30}$$

which we can employ to decrease calculation cost. Suppose we have N images to be evaluated, which means there are $N - 1$ relative quality indices to be calculated. For each index, $N - 1$ kernel functions should be computed, resulting in the cost of $(N - 1)^2$ iterations. Figure 8.17 visualizes the iterations when $N = 5$. The red cross indicates no need of calculation, and green patch denotes calculating the kernel function with an image at the row and the other image at the column.

When the symmetry is met, one-half of iterations are reduced. Figure 8.18 illustrates the relevant procedures where the gray and black arrows indicate previous and current reduced iterations, respectively.

8.4 Adaptive Paralleled Sinogram Noise Reduction for Low-Dose X-Ray CT Based on Relative Quality Assessment

8.4.1 Introduction

X-ray computed tomography (CT) is an important and useful clinical tool in assisting diagnosis and treatment. In recent years, reconstruction of low-dose X-ray CT has become an advanced research hotspot. However, the artifacts caused by physical effects, e.g., noise, misalignment, and scanner effects, e.g., scattering,

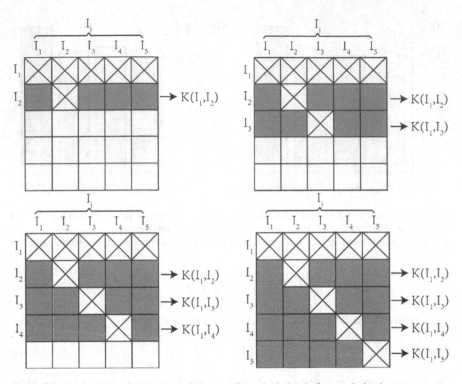

Fig. 8.17 Procedure on five images relative quality calculation before optimization

beam hardening, would affect the diagnosis accuracy (Karimi et al. 2016). High signal-to-noise ratio of CT image can only be achieved with high dose. In other words, noise is a core issue affecting quality of low-dose CT images (Xu et al. 2012; Othman et al. 2016; Zhu and Ding 2017). In the literature, many methods have been proposed to figure these effects out and improve the quality of noisy CT images.

Such denoising methods can be roughly divided into two parts. One part of them creates a noise property in image domain and minimizes the cost function by iterative algorithms (Baumueller et al. 2012; Schuhbaeck et al. 2013; Yan et al. 2014). For example, Elbakri et al. discovered that the detected photon numbers were considered following a Poisson distribution plus a background Gaussian noise with a zero mean and proposed a statistical iterative reconstruction algorithm (Elbakri and Fessler 2002). Maximum a posteriori (MAP) estimation method and iterated conditional mode (ICM) algorithm were used to smooth sinogram non-linearly (Li et al. 2004). The other part attempts to reduce noise in sinogram domain by modeling the noise property and conducting the Radon transformation by a standard filtered backprojection reconstruction (FBP) algorithm or a Feldkamp–Davis–Kress (FDK) method (Feldkamp et al. 1984; Manduca et al. 2009; Ehman et al. 2012; Li et al. 2014). As an example, the Gauss–Seidel Penalized Reweighted

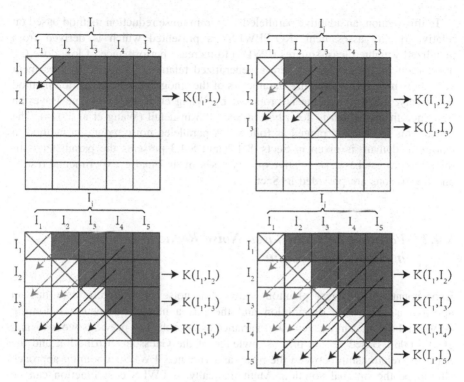

Fig. 8.18 Procedure on five images relative quality calculation using symmetrical optimization

Least-Squares (GS-PRWLS) method aims to estimate the ideal sinogram and reconstruct CT images using a standard FBP algorithm (Wang et al. 2006). Generally, the statistical methods in sinogram domain are more efficient and simpler than statistical reconstruction through image domain (Manduca et al. 2009). However, because of the huge amount of calculation resulting from the large quantity of sinogram data, the efficiency of those denoising algorithms is still unacceptable.

For fast computation on multiprocessor platform, especially graphics processing unit (GPU) accelerator, paralleled methods are studied by many researchers (Gao 2012; Li et al. 2014; Yan et al. 2014; Matenine et al. 2015; Xu et al. 2016). Nevertheless, those paralleled methods still cannot be able to apply in the time-critical environment due to long computational time. For example, the NCG-type method proposed by Jeffrey A. Fessler et al. is easier to be parallelized because they update all voxels simultaneously (Fessler and Booth 1999). Later, an adaptive nonlocal means (NLM) filter based on local noise level was developed and implemented on GPU, which significantly cut down the computational time (Li et al. 2014). However, these methods still cost too much resource to complete sinogram noise reduction and more or less need to set some parameters so that they still cannot be applied in the time-critical environment.

In this section, an adaptive paralleled sinogram noise reduction method based on relative quality assessment (RQ-PPWLS) is presented which is derived from penalized weighted least-squares (PWLS) to increase performance of low-dose CT noise reduction algorithms, where a generalized relative quality (RQ) assessment scheme is involved to optimize parameters of the sinogram noise reduction method adaptively. The RQ scheme is proposed by Wang et al. based on the Bayesian inference theory which is described in Sect. 8.3 in detail (Wang et al. 2014). The rest of this section is arranged as follows. A paralleled noise reduction method in sinogram domain is given in Sects. 8.4.2 and 8.4.3 presents the parameter optimization process based on relative quality assessment. Finally, experimental results and discussions are provided in Sect. 8.4.4.

8.4.2 Paralleled PWLS-Based Noise Reduction in Sinogram Domain

Experiments of physical phantom show that noise of sinogram data has an approximate Gaussian distribution and there is a nearly exponent relationship between the data sample mean and variance as described in previous work (Wang et al. 2006). Based on the prior knowledge of the Gaussian distribution and the exponent relationship between the mean and variance, PWLS smoothing approach should be the optimal solution. Mathematically, a PWLS cost function can be described as follows

$$\phi(q) = (\hat{y} - q)' \sum{}^{-1}(\hat{y} - q) + \beta R(q) \tag{8.31}$$

where $\hat{y}$ is the vector of line integral after system calibrating and logarithm transform, q is the vector of ideal projection data, Σ^{-1} is diagonal matrix of variance, the first term is the Weighted Least-Squares (WLS) cost function. And β is a smooth parameter controlling the proportion between the ideal data and the line integral, $R(q)$ is the penalty term,

$$R(q) = q'Rq = \frac{1}{2}\sum_{i}\sum_{m \in Ni} \omega_{im}(q_i - q_m)^2 \tag{8.32}$$

where N_i indicates the set of first-order neighbors of the ith pixel in the sonogram, and ω_{im} is the weight along horizontal and vertical directions.

The PWLS method is aimed to minimize Eq. (8.31) by estimating q from $\hat{y}$,

$$q = \arg \min_{q \geq 0} \phi(q) \tag{8.33}$$

To solve this problem, numerical classical methods, such as Gauss–Seidel iteration, fail to update all pixels of sinogram in one iteration, which results in a low computational efficiency, because pixels are correlative. In contrast, this book removes correlation between pixels by finding the cross-zero point of the partial derivative of Eq. (8.31) to improve the efficiency,

$$\frac{\partial \phi(q)}{\partial q_i} = 2(q_i - \hat{y}_i) \sum{}^{-1} + \beta \sum_i \sum_{m \in Ni} \omega_{im}(q_i - q_m) \qquad (8.34)$$

Therefore, a pixel-free and paralleled iterative formula can be written as,

$$q_i^{(n+1)} = \frac{2\sum{}^{-1} q^{(n)} + \beta \sum_{m \in Ni} \omega_{im} q_m^{(n)}}{2\sum{}^{-1} + \beta \sum_{m \in Ni} \omega_{im}} \qquad (8.35)$$

For Σ^{-1} calculation, the variance of the repeated measurements of projections (after system calibration and log transform) σ^2 is defined as

$$\sigma_i^2 = f_i \times \exp(q_i/\eta) \qquad (8.36)$$

where q_i is the logarithm transform of projection data (line integral), η is a scaling parameter, and f_i is a parameter based on different detector bins (Wang et al. 2006).

Obviously, in Eq. (8.35), the ith pixel in q is only related to itself and does not rely on iterative results from other pixels. Consequently, this simple iterative formula can be easily paralleled and deployed on multiprocessor platform to obtain a high computational efficiency. Implementation of the paralleled PWLS-based sinogram noise reduction method for the solution of Eq. (8.35) is summarized in Table 8.5. Step (1) is the initialization, where after system calibration, projection data is transformed to line integral by logarithm transform. In step (2), variance is calculated from line integral based on exponent relationship. Then according to Eq. (8.35), q is updated for noise reduction. Here, iterations are determined by relative image quality assessment. Finally, in the last step, CT image is reconstructed from the denoised data for output.

Table 8.5 Procedure of paralleled PWLS sinogram noise reduction

Step	(1)	System calibration and logarithm transform of projection data (line integral)
	(2)	Calculate the variance σ^2 based on Eq. (8.36)
	(3)	Update all pixels of line integral paralleled according to (5)
	(4)	Repeat Step (2) and Step (3) until the best relative quality of sinogram data is got
	(5)	Output a high-quality CT image

8.4.3   Automatic Parameter Optimization Based on Relative Quality

In paralleled PWLS algorithm, different parameter β generates reconstructed images with different quality. To optimize the estimation of sinogram, it is necessary to evaluate the objective relative quality of these denoised sinogram with different smooth parameter β. In the proposed method, for parameter optimization, a generalized relative quality (RQ) is utilized to evaluate each relative quality of sinogram denoised by different smooth parameters.

Given N groups of parameters, the paralleled PWLS algorithm generates different sonogram, i.e., $\{S_1, S_2, ..., S_N\}$. In the RQ calculation, every sinogram can be chosen as reference and others are regarded as distorted. Thus, the relative quality of the distorted sinogram can be assessed by a full-reference metric. In this way, from $\{S_1, S_2, ..., S_N\}$, the relative quality defined as $\{RQ(S_1), RQ(S_2), ..., RQ(S_N)\}$ are calculated. One by one, with all of the different sinogram data being considered as reference there are N RQ values as a result. Comparing them to get the best one, a relative quality assessment process is accomplished. Mathematically, the expression of RQ can be written as

$$RQ(S_i) = \sum_{j \neq i}^{N} F_{IQA}(S_i, S_j) \tag{8.37}$$

where S_i is marked as a reference sinogram and S_j is the distorted one. F_{IQA} is a full-reference image quality assessment metric, such as peak signal-to-noise ratio, structure similarity index metric (Wang et al. 2004). For simplicity, peak signal-to-noise ratio (PSNR) is used to evaluate each relative quality of sinogram denoised by different smooth parameters.

$$RQ_{PSNR} = 20 \log \frac{255}{\sqrt{\frac{1}{MN} \sum_{x=1}^{N} \sum_{y=1}^{M} [I_{dis}(x, y) - I_{ref}(x, y)]^2}} \tag{8.38}$$

where M and N are the height and width of the sinogram, respectively, $I_{dis}(x, y)$ is a pixel of distorted sinogram located on (x, y), and $I_{ref}(x, y)$ represents the corresponding pixel in reference sinogram.

Figure 8.19 illustrates the scheme of parameter optimization by using relative quality assessment. With all the sinogram data being handled in this scheme, the optimized parameter is selected.

Finally, the entire procedure of the proposed adaptive paralleled sinogram noise reduction and reconstruction method (RQ-PPWLS) is shown in Fig. 8.20. At first, after system calibration and logarithm transformation of sinogram data, the line integral is put into the module of RQ-based parameter optimization. Then the paralleled module generates several different outputs with a range of parameters and utilizes RQ assessment method to obtain relevant results. Next, the highest RQ

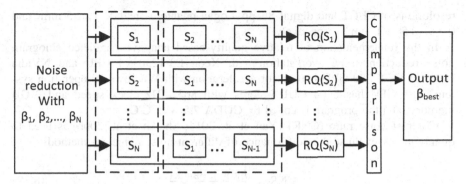

Fig. 8.19 Scheme of parameter optimization

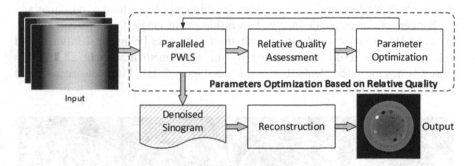

Fig. 8.20 Procedure of adaptive parraleled sinogram noise reduction and reconstruction

value for parameter optimization is chosen and fed back to the paralleled step. Finally, a CT image with high quality can be reconstructed by FDK directly from the denoised sinogram data.

8.4.4 Experimental Results and Discussions

Experiments are conducted on a modified Shepp–Logan digital phantom (Shepp and Logan 1974) and two practical sets of cone beam CT (CBCT) sinogram data. The sinogram data of Shepp–Logan is calculated from Siddon's forward projection algorithm (Siddon 1985), which has 656 projections in a 360-degree full-fan scan. The first CBCT set is based on a Catphan phantom, which has 655 projections in a 360-degree full-fan scan. The other comes from an anonymous patient data, which has 656 projections in 360-degree half-fan scan. Each projection is composed of 1024×768 pixels with a resolution of 0.388×0.388 mm^2. The volume of three reconstructed images has a size of $512 \times 512 \times 32$ for testing, while the voxel resolutions of CBCT and digital Shepp–Logan phantom are $1 \times 1 \times 2.5$ mm^3 and $0.5 \times 0.5 \times 1$ mm^3, respectively.

In the implementation of relative quality-based adaptive paralleled sinogram noise reduction, a PC workstation with Xeon E3-1230 v3 CPU and NVidia GeForce GTX 780 GPU is used for acceleration with Windows 10 operation system. The GPU has 2304 CUDA cores with 863 MHz clock speed and 3 GB memory, and the program is coded by CUDA 7.5 with C/C++.

Contrast–noise ratio (CNR) (Yan et al. 2014; Sheikh et al. 2006) is used to quantitatively evaluate the performance of different noise reduction methods.

$$\text{CNR} = \frac{|\mu_{\text{ROI}} - \mu_{\text{BACK}}|}{\sigma_{\text{ROI}} + \sigma_{\text{BACK}}} \tag{8.39}$$

where μ and σ represent the mean and standard deviation. Experimental results of modified Shepp–Logan digital phantom are illustrated in Fig. 8.21, where zoomed-in ROIs (regions of interest) are highlighted inside the dashed circle and the background areas are defined as that inside the rectangle for CNR calculation. Figure 8.21a is the ideal phantom with the CNR value being ideal infinity, and Fig. 8.21e is the result reconstructed by the FDK method without adding any noises, which means an ideal CT scan process. Figure 8.21b, f is reconstructed from

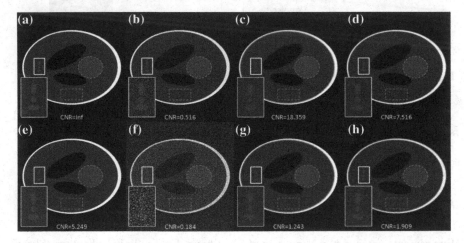

Fig. 8.21 Experiment results on modified Shepp–Logan digital phantom. A zoomed-in ROIs are shown in lower-left corner, and the areas in dashed circle and rectangle are defined as ROI and background, respectively, to calculate CNR values. **a** Ideal modified Shepp–Logan phantom. **b** Projections are added Gaussian white noise with SNR = 20 dB then reconstructed by FDK without noise reduction method. **c** Gaussian white noise of SNR = 20 dB, denoised and reconstructed by GS-PRWLS. **d** Gaussian white noise of SNR = 20 dB, denoised and reconstructed by RQ-PPWLS. **e** After Siddon's forward projection without adding noise then reconstructed. **f** Gaussian white noise of SNR = 10 dB, without noise reduction. **g** Gaussian white noise of SNR = 10 dB, denoised and reconstructed by GS-PRWLS. **h** Gaussian white noise of SNR = 10 dB, denoised and reconstructed by RQ-PPWLS

the same data set used in Fig. 8.21e with Gaussian white noise of SNR = 20 dB and SNR = 10 dB, respectively. Obviously, details shown in zoomed-in ROI are lost in the condition of SNR = 10 dB (CNR = 0.184) while some useful information is barely visible under disturbance from 20 dB white noise (CNR = 0.516). Figure 8.21c displays the result by GS-PRWLS (Wang et al. 2006) noise reduction algorithm, while the results of RQ-PPWLS are in Fig. 8.21d. Similarly, Fig. 8.21f–h are the results of different denoising methods in the case of Gaussian white noise of SNR = 10 dB.

From the comparison of Fig. 8.21c–e, we can find that: (i) Both GS-PRWLS and the presented RQ-PPWLS successfully reduce the noise from the original sinogram in the full-fan mode. (ii) Subjectively, the result generated from RQ-PPWLS has the better performance in preserving the image edge, while the denoised image by GS-PRWLS in Fig. 8.21c is over blurred. (iii) Although the CNR value of GS-PRWLS (18.359) is higher than that of RQ-PPWLS (7.516), the overall subjective evaluation on Fig. 8.21d is more positive than that on Fig. 8.21c. The reason lies in that CNR is not a suitable metric for medical image quality assessment. That is to say, it is desirable to develop an effective medical image quality assessment method that is consistent with subjective evaluation. As another case of study, the comparison of Fig. 8.21g, h reveals that the image denoised and reconstructed by the proposed RQ-PPWLS method has higher CNR value (1.909) than that of GS-PRWLS (1.243), and that Fig. 8.21h contains more detail information while there are some streaking artifacts in Fig. 8.21g.

The reconstructed results of Catphan phantom are shown in Fig. 8.22. The CNR criteria indicate the superiority of both GS-PRWLS and RQ-PPWLS over FDK method. As shown in the zoomed-in ROIs, RQ-PPWLS has a better subjective quality than GS-PRWLS; meanwhile, the CNR value is increased from 1.4158 (FDK) through 2.1102 (GS-PRWLS) to 2.4602 (RQ-PPWLS), indicating that the proposed RQ-PPWLS method achieves a better performance of denoising.

The anonymous patient case is shown in Fig. 8.23. Also, there is a zoomed-in ROI within a box which is placed in the upper-right corner, and the areas inside the dashed box are defined as background, while the highlight area is defined as ROI. Obviously, the result from FDK method has many artifacts with a low image quality. After noise reduction processing, in contrast, RQ-PPWLS removes these artifacts satisfactorily while result from GS-PRWLS still has a light wide stripe across middle of the CT image. As a result, in the zoomed-in ROIs, RQ-PPWLS method has the peak CNR value (1.4374) which is significantly promoted from direct FDK reconstruction (0.5957) and also better than GS-PRWLS method (1.2245).

In addition, the comparison of computation time is reported in Table 8.6. The proposed RQ-PPWLS accelerates the current GS-PRWLS noise reduction procedure by a rate of 323.9 in Catphan mode and of 244 in patient case. In addition, RQ-PPWLS processes 655 projections of Catphan phantom in 8.15 s and 656 projections of patients in 10.2 s, which means the proposed method can also be applied in practice scene. Because this method updates all voxels simultaneously, theoretically, it can be faster by loading all projections if there were enough

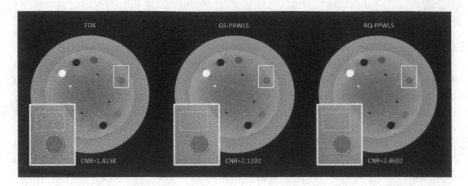

Fig. 8.22 Reconstructed images of Catphan phantom (full-fan mode). Contrast slice is shown with windows (0.0115, 0.0245). From left to right: directly FDK reconstruction; sinogram denoised by GS-PRWLS then FDK reconstruction; sinogram denoised by RQ-PPWLS then FDK reconstruction

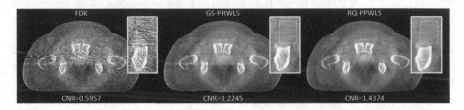

Fig. 8.23 Reconstructed images of anonymous patient (half-fan mode). CT image is shown with windows (0.0115, 0.0245). From left to right: directly FDK reconstruction, sinogram denoised by GS-PRWLS then FDK reconstruction, sinogram denoised by RQ-PPWLS then FDK reconstruction

memory in the workstation system. Still, there are many reasons to limit the operating speed. On the one hand, owing to lacking enough GPU memory, RQ-PPWLS only processes 16 projections synchronously. On the other hand, the shortage of GPU memory causes many I/O operations, which also slow the entire procedure. In addition, the experiments are operated on a windows system with GUI, which means GPU accelerator cannot be utilized in maximum efficiency. Experiments validate that RQ-PPWLS has the potential to speed up by solving all these problems above.

Furthermore, Fig. 8.24 shows the results of relative quality assessment for sinogram denoised with different parameters where ω_{im} represents the horizontal to vertical ratio range in Eq. (8.32) and β is the smooth parameter in Eqs. (8.31), (8.34), and (8.35). In this case, the worst RQ value is defined as 0 while the best is 100. Then it can be directly observed that the value of relative quality reaches the peak when ω_{im} is close to 0.67 and β equals to 1100. According to the result of relative quality assessment, the optimized parameters are confirmed and delivered to the following steps in the denoising scheme. It confirms that the RQ-PPWLS has the ability of adaptive optimization and achieving a good performance of noise reduction. It should be noted that in the relative quality assessment, PSNR is applied as quality evaluation kernel algorithm because of its efficiency and convenience. Since PSNR is often blamed for its inaccuracy and inconsistency with subjective evaluation without considering the characteristics of human visual system (HVS) (Zhang and Chandler 2013). In future work, other better performance image quality methods can be employed as RQ assessment kernel instead of PSNR.

In summary, a paralleled iterative method for PWLS is developed by updating all pixels simultaneously, which makes a remarkable progress on reduction of computational time. Also, quality of denoised image is improved noticeably by relative image quality assessment-based parameter optimization.

8.5 Study on the Relationship Between the Image Quality and Imaging Dose in Low-Dose CBCT

8.5.1 Introduction

The issue of radiation dose exposure to patients from digital radiography is a major public health concern (Gonzalez and Darby 2004; Brenner and Hall 2007; Xu et al. 2012). Recently, with the wide usage of flat panel X-ray cone beam CT (CBCT) imaging in image guided radiotherapy (IGRT) (Jaffray et al. 2002; McBain et al. 2006; Grills et al. 2008), a considerable amount of excessive radiation dose to radiotherapy patients (Daly et al. 2006; Islam et al. 2006; Ding and Coffey 2009) has drawn more and more attention (Brenner et al. 2001). However, the image quality will be degraded inevitably when the imaging dose decreases. It is well known that there is a trade-off between noise level and radiation dose (Deng et al. 2015). On the one hand, high-dose radiation will lower the noise level. On the other hand, low-dose radiation will lower the SNR of the image and result in reducing the amount of image information. Therefore, it is necessary to make a clear understanding of the relationship between the image quality and imaging dose at low-dose levels (Kim et al. 2010; Han et al. 2015; Bian et al. 2016). And based on such understanding, it is desirable to design optimal low-dose scan protocols to maximize the dose reduction while minimizing the image quality loss for various specific IGRT tasks (Yan et al. 2012a, b).

In the literature, some preliminary work has been done. For example, Tang et al. claimed that it is preferable to distribute the total imaging dose into many view angles in order to reduce the under-sampling streaking artifacts (Tang et al. 2009). Han et al. compared three different dose allocation schemes at a constant dose level (Han et al. 2010). However, both studies are based on a single-dose level. Later, Yan et al. presented a comprehensive study on the relationship between the image quality and imaging dose under the CS reconstruction framework (Yan et al. 2012a, b). In their distinguished work, a dose-quality map (DQM) is generated

Table 8.6 Time cost by different methods

Algorithm	Catphan Phantom[a]		Patient[a]		Other		
	GS-PRWLS	RQ-PPWLS	GS-PRWLS	RQ-PPWLS	NLM[b]	Bilateral filtering[c]	TV-based filtering[c]
Projections	655		656		300	240	
Resolution	1024 × 768					875 × 568	
Time cost	2640 s[d]	8.15 s[d]	2489 s[d]	10.2 s[d]	5 min	8.5 min	6 min
Speedup	323.9		244		/	/	/

[a] Operation on Intel Xeon E3 1230 V3, one NVidia GeForce GTX 780 with 16 GB Memory system in Windows 10 system

[b] Operation on Intel Xeon CPU cores running at 2.4 GHz with 24 G RAM, and 8 NVidia GeForce GTX 680 (Li et al. 2014)

[c] Operation on a Windows 7 PC with 16 GB of memory and 3.4 GHz Intel Core i7 CPU (Karimi et al. 2016)

[d] Time cost only for sinogram noise reduction, not include time of reconstruction

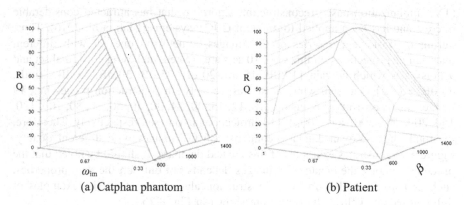

(a) Catphan phantom   (b) Patient

Fig. 8.24 Relative Quality of different parameters: **a** RQ value calculated during RQ assessment in Catphan phantom, **b** RQ value of patient case

where image quality and imaging dose are modelled as functions of two variables, i.e., the number of projections and mAs per projection (mAs/view). With such DQM, we can have a clear global picture of the trade-off between the image quality and imaging dose.

In this section, we study on the relationship between the image quality versus imaging dose and noise levels under different reconstruction framework comprehensively based on dose-quality maps.

8.5.2   Medical Image Databases Creation

Generally speaking, low-dose scanning can be achieved by either fixing the number of projections while decreasing the X-ray tube load (mAs) or fixing the mAs level while decreasing the number of projections (Yan et al. 2012a, b). In experiments, we create three databases based on Shepp–Logan digital phantom (Shepp and Logan 1974), Catphan phantom, and real data of patients, respectively. All the original data is processed by different methods to generate CT images with different quality, such as adding Gaussian white noise, reducing the number of projections.

In the procedure of Shepp–Logan database's creation, an image matrix with resolution of $512 \times 512 \times 512$ is firstly generated by MATLAB phantom function as a reference CT image. Then the forward projection method (Siddon 1985) is utilized to produce 1200 projections in different angles which are defined as the original projection set. Next, different levels of Gaussian white noise are added to the original set to create different noisy projection sets to simulate low-dose CT imaging in practice. Finally, all of projection sets are sent to reconstruction program to calculate CT images with noise and artifacts. The reconstruction algorithms include FDK (Feldkamp et al. 1984), GS-PRWLS (Wang et al. 2006) and ASD-POCS (Sidky et al. 2011), where ASD-POCS is a total variation

(TV) minimization-based reconstruction algorithm that has attracted considerable interest and has been applied to different CBCT systems (Bian et al. 2016). Thus, during the last step, a series of CT images come from one set with different quantitative projections. In this way, 10 sets are created, including 1 original set and 9 noisy sets which contain {10 dB, 15 dB, 20 dB, 25 dB, 30 dB, 35 dB, 40 dB, 45 dB, 50 dB} Gaussian white noise, respectively. In construction processing, 19 images are reconstructed by using {10, 12, 15, 20, 24, 30, 40, 50, 60, 80, 100, 120, 150, 200, 240, 300, 400, 600, 1200} projections in one set, respectively. Therefore, the total Shepp–Logan database contains 486 CT images with different quality. Figure 8.25 illustrates examples of this created database. It should be noted that the difference among the created CT images depends not only on the data properties, such as noise, but also on the reconstruction algorithms used. The examples of different reconstruction algorithms are shown in Fig. 8.26

In the Catphan phantom database creation, there are 13 sets containing 655 projections with different CT scan parameters whose current are set to {5 mA, 10 mA, 15 mA, 20 mA, 25 mA, 30 mA, 35 mA, 40 mA, 45 mA, 50 mA, 60 mA, 70 mA, 80 mA}, respectively, while voltage is fixed to 125 kV. Just like the same reconstruction strategy applied in Shepp–Logan database, 16 images are outputted by the construction program, using {10, 13, 16, 21, 27, 32, 43, 54, 65, 81, 109, 131, 163, 218, 327, 655} projections in one set. In other words, there are 624 images in Catphan phantom, examples of which are shown in Fig. 8.27. Additionally, the examples of different reconstruction algorithms are shown in Fig. 8.28.

In the database creation based on real data of patients, there are seven pelvis cone beam CT (CBCT) half-fan scan raw data and five planning CT

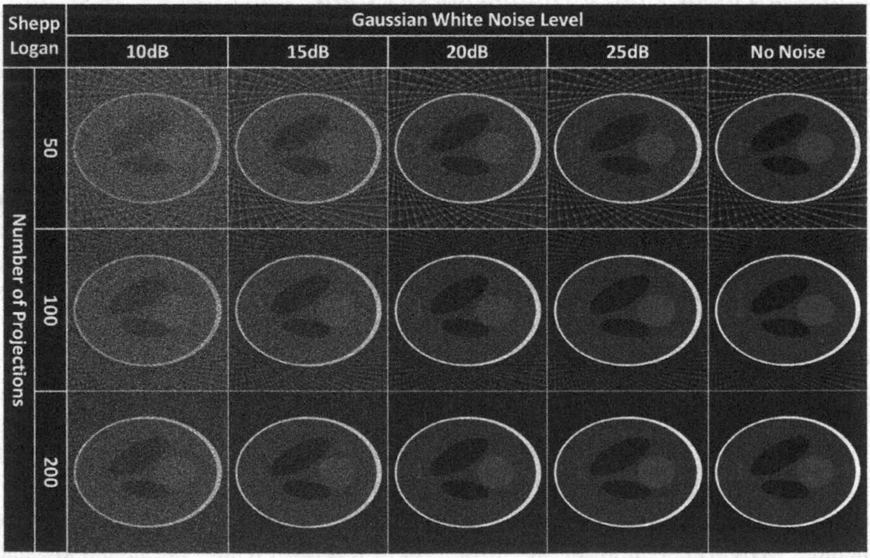

Fig. 8.25 Examples of Shepp–Logan phantom database

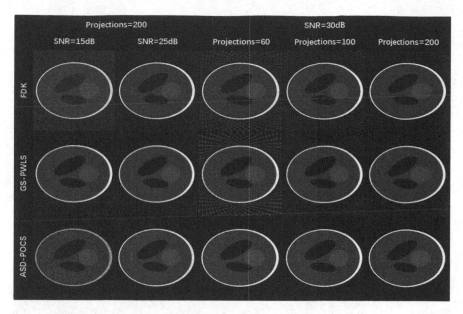

Fig. 8.26 Examples of different reconstruction algorithms on Shepp–Logan phantom database

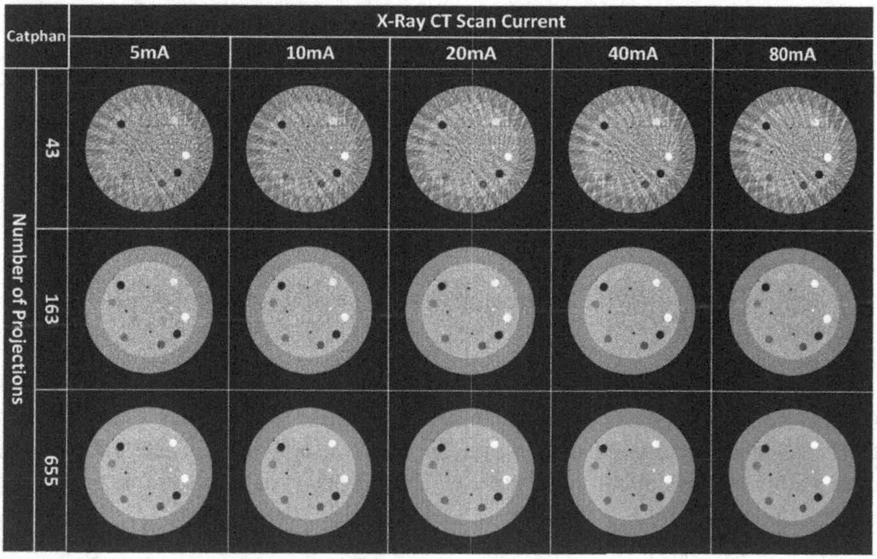

Fig. 8.27 Examples of Catphan phantom database

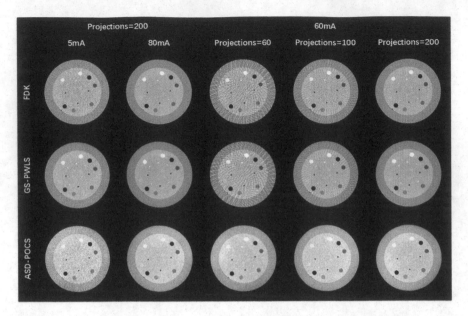

Fig. 8.28 Examples of different reconstruction algorithms on Catphan phantom database

Table 8.7 Details of patient database

Patient	1	2	3	4#1	4#2	5#1	5#2
CBCT projections	654	662	655	656	654	655	657
CBCT images	16 (Level of quality) × 10 (Number of images in one quality level) = 160						
CBCT voxel size (mm)	0.8 × 0.8 × 2.5						
pCT images	58	61	58	56		88	
pCT voxel size (mm)	1.15 × 1.15 × 5	1.16 × 1.16 × 5	0.98 × 0.98 × 5	1.46 × 1.46 × 5		1.11 × 1.11 × 3	

(pCT) reconstructed data of five patients. Among these seven CBCT scans, the first three patients are scanned once while the last two are twice. Table 8.7 lists the detailed information about this database. Thus, 7 pelvis data sets are reconstructed using {1/60, 1/50, 1/40, 1/30, 1/24, 1/20, 1/15, 1/12, 1/10, 1/8, 1/6, 1/5, 1/4, 1/3, 1/2, 1} of CBCT projections to obtain 16 image sets (contain 10 images) with different quality level. In addition, parameters of CBCT scan are set to 125 kV and 80 mA. Also, examples are displayed in Fig. 8.29.

For the CNR calculation, the definition of the ROI and background areas is shown in Fig. 8.30.

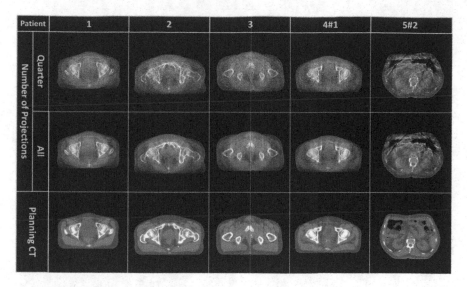

Fig. 8.29 Examples of patient database

8.5.3   Medical Image Quality Assessment Metrics

For the experimental data, the CNR is used to investigate the visibility of the ROIs shown in Fig. 8.30, and noise level (NL) is used to measure the noise level in the medical images. The calculation of CNR is shown above as Eq. (8.39). And NL is calculated as following,

$$\text{NL} = \frac{1}{MN} \sum \frac{|I(x,y) - \mu_{\text{block}}|}{\sigma^2_{\text{block}}} \tag{8.40}$$

where $I(x, y)$ is a pixel of the medical image, M and N are the height and width of the neighboring block, μ_{block} and σ_{block} are the mean value and variance of the neighboring block, respectively.

And to quantify the medical image quality quantitatively, PSNR, SSIM (Wang et al. 2004), and a method with general statistic features extraction based on topographic independent component analysis (TICA) (Ding et al. 2014) are used as the quality assessment metrics.

Certainly, current metrics for medical image quality assessment are hardly consistent with subjective evaluation very well. Each metric is more or less biased by emphasizing certain aspects of the image quality. However, objective metrics can help identify the best image with a quantitative score.

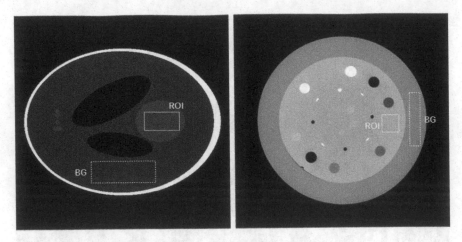

Fig. 8.30 Setting of ROI and BG regions in Shepp–Logan and Catphan

8.5.4   Experimental Results and Discussion

Under different reconstruction algorithms, DQM, as a function of the number of projections and mAs per projection, can visualize the relationship between the image quality and imaging dose or noise level (Yan et al. 2012a, b). DQMs in terms of CNR, PSNR, SSIM, NL, and TICA on the created Shepp–Logan database are shown in Figs. 8.31a–e, where the x-axis is the noise level ranging from 10 to 50 dB, the y-axis is the number of projections ranging from 10 to 600, and the color represents the IQA scores.

From Fig. 8.31, we can observe that (1) image quality degradation becomes evident when the noise level decreases below 35 dB. (2) In contrast to the FDK algorithm, GS-PRWLS and ASD-POCS are able to handle both the sparse-view reconstruction and low-dB reconstruction, achieving comparable image quality. (3) For the case of very low noise level, reasonable image quality can be achieved with a large amount of views.

Figure 8.32 shows the DQMs in terms of CNR, NL, and TICA on the created Catphan database, where the x-axis is the mAs level ranging from 5 to 80 mA, the y-axis is the number of projections ranging from 10 to 655, and the color represents the IQA scores. From Fig. 8.32, we can find that:

1. Image quality has little degradation over a large range of imaging dose variation (upper-right corner), but changes rapidly at the low-dose range (lower-left corner).
2. Image quality varies slowly over a large imaging dose range above 70 mAs, more quickly when below 30 mAs.
3. It is a challenging case that the super sparse-view reconstruction with the small projection number no matter what kind of scan protocols. For the case of

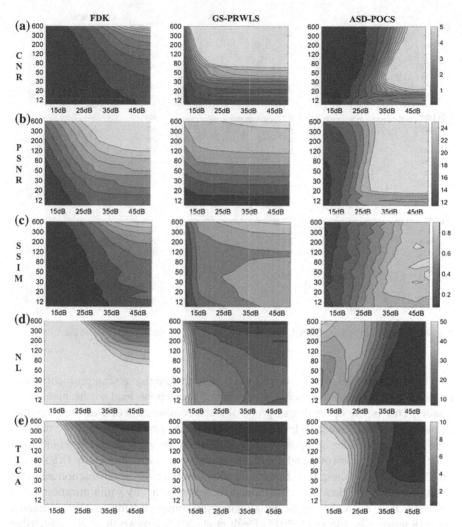

Fig. 8.31 DQMs in terms of CNR, PSNR, SSIM, NL, and TICA on the created Shepp–Logan database

extremely sparse views (projection number <50), image quality degrades severely no matter how high the mAs level is.

4. For the case of very low mAs level, reasonable image quality can be achieved with a large amount of views.

5. Compared with the classic FDK algorithm, CS-based reconstruction algorithms can handle both the sparse-view reconstruction and low-mAs reconstruction, achieving acceptable image quality. These observations quite agree with those in (Yan et al. 2012a, b). However, at present, it should be noted that the clinically acceptable lowest imaging dose level is task dependent.

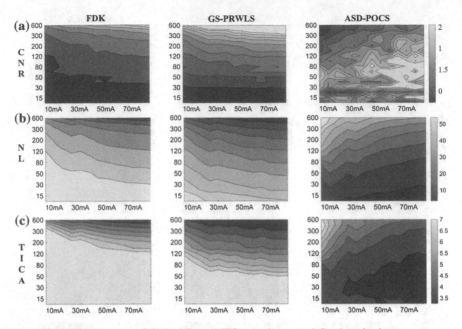

Fig. 8.32 DQMs in terms of CNR, NL, and TICA on the created Catphan database

The results of Figs. 8.31 and 8.32 explicitly indicate that a scan protocol with a medium mAs level and a medium number of projections lead to the best image quality. That is, the optimal scan protocol is the combination of a medium number of projections and a medium level of mAs. The underlying reason might be that this scan protocol provides an optimal balance between the view sampling requirement under the CS framework and the SNR demand in each projection (Yan et al. 2012a, b). Besides, there is a strong indication that iterative reconstruction methods from low-dose CT data and constrained total variation (TV) minimization-based reconstruction methods can potentially improve image quality or lower imaging dose (Ramirez-Giraldo et al. 2011; Deák et al. 2013; Hoxworth et al. 2014; Bian et al. 2016).

References

Acharya, T., & Ray, A. K. (2005). Image processing: Principles and applications. Wiley.
Baumueller, S., Winklehner, A., Karlo, C., Goetti, R., Flohr, T., Russi, E. W., et al. (2012). Low-dose CT of the lung: Potential value of iterative reconstructions. *European Radiology, 22* (12), 2597–2606.
Beyersdorff, D., Taymoorian, K., Knösel, T., Schnorr, D., Felix, R., Hamm, B., et al. (2005). MRI of prostate cancer at 1.5 and 3.0 T: Comparison of image quality in tumor detection and staging. *American Journal of Roentgenology, 185*(5), 1214–1220.

Bian, J., Sharp, G. C., Park, Y., Ouyang, J., Bortfeld, T., & Fakhri, G. E. (2016). Investigation of cone-beam CT image quality trade-off for image-guided radiation therapy. *Physics in Medicine & Biology, 61*(9), 3317–3346.

Bohning, D. E., Lomarev, M., Denslow, S., Nahas, Z., Shastri, A., & George, M. (2001). Feasibility of vagus nerve stimulation–synchronized blood oxygenation level–dependent functional MRI. *Investigative Radiology, 36*(8), 470–479.

Brenner, D. J., Elliston, C. D., Hall, E. J., & Berdon, W. E. (2001). Estimated risks of radiation-induced fatal cancer from pediatric CT. *American Journal of Roentgenology, 176*(2), 289–296.

Brenner, D. J., & Hall, E. J. (2007). Computed tomography—An increasing source of radiation exposure. *The New England Journal of Medicine, 357*(22), 2277–2284.

Cavaro-Ménard, C., Zhang, L., & Callet, P. L. (2010). Diagnostic quality assessment of medical images: Challenges and trends. In *2nd European Workshop on Visual Information Processing, Paris, France*. Piscataway, USA: IEEE, pp. 277–284.

Chang, C. C., & Lin, C. J. (2011). LIBSVM: a library for support vector machines. *ACM Transactions on Intelligent Systems and Technology (TIST), 2*(3), 1–27.

Chapman, D., Thomlinson, W., Johnston, R. E., Washburn, D., Pisano, E., Gmür, N., et al. (1997). Diffraction enhanced x-ray imaging. *Physics in Medicine & Biology, 42*(11), 2015–2025.

Cortes, C., & Vapnik, V. (1995). Support-vector networks. *Machine Learning, 20*(3), 273–297.

Cosman, P. C., Gray, R. M., & Olshen, R. A. (1994). Evaluating quality of compressed medical images: SNR, subjective rating, and diagnostic accuracy. *Proceedings of the IEEE, 82*(6), 919–932.

Cunningham, P. M., Brennan, D., O'Connell, M., Macmahon, P., O'Neill, P., & Eustace, S. (2007). Patterns of bone and soft-tissue injury at the symphysis pubis in soccer players: Observations at MRI. *American Journal of Roentgenology, 188*(3), W291–W296.

Daly, M., Siewerdsen, J., Moseley, D., Jaffray, D., & Irish, J. (2006). Intraoperative cone-beam CT for guidance of head and neck surgery: Assessment of dose and image quality using a C-arm prototype. *Medical Physics, 33*(10), 3767–3780.

Deák, Z., Grimm, J. M., Treitl, M., Geyer, L. L., Linsenmaier, U., Körner, M., et al. (2013). Filtered back projection, adaptive statistical iterative reconstruction, and a model-based iterative reconstruction in abdominal CT: An experimental clinical study. *Radiology, 266*(1), 197–206.

Deng, C., Ma, L., Lin, W., & Ngan, K. N. (2015). *Visual signal quality assessment*. Switzerland: Springer International Publishing.

Denk, C., & Rauscher, A. (2010). Susceptibility weighted imaging with multiple echoes. *Journal of Magnetic Resonance Imaging, 31*(1), 185–191.

Dias, J. M. P., Oliveira, C. M., & Cruz, L. A. D. S. (2014). Retinal image quality assessment using generic image quality indicators. *Information Fusion, 19*(1), 73–90.

Ding, G. X., & Coffey, C. W. (2009). Radiation dose from kilovoltage cone beam computed tomography in an image-guided radiotherapy procedure. *International Journal of Radiation Oncology Biology Physics, 73*(2), 610–617.

Ding, Y., Dai, H., & Wang, S. Z. (2014). Image quality assessment scheme with topographic independent components analysis for sparse feature extraction. *Electronics Letters, 50*(7), 509–510.

Dobbin, J. T., III, Samei, E., Ranger, N. T., & Chen, Y. (2006). Intercomparison of methods for image quality characterization. II. Noise power spectrum. *Medical Physics, 33*(5), 1466–1475.

Ehman, E. C., Guimarães, L. S., Fidler, J. L., Takahashi, N., Ramirez-Giraldo, J. C., Yu, L., et al. (2012). Noise reduction to decrease radiation dose and improve conspicuity of hepatic lesions at contrast-enhanced 80-kV hepatic CT using projection space denoising. *American Journal of Roentgenology, 198*(2), 405–411.

Elbakri, I. A., & Fessler, J. A. (2002). Statistical image reconstruction for polyenergetic X-ray computed tomography. *IEEE Transactions on Medical Imaging, 21*(2), 89–99.

Fasih, M., Langlois, J. M. P., Tahar, H. B., & Cheriet, F. (2014). Retinal image quality assessment using generic features. In *Proceedings of SPIE* (Vol. 9035, pp. 90352Z).

Feldkamp, L., Davis, L., & Kress, J. (1984). Practical cone-beam algorithm. *Journal of the Optical Society of America A, 1*(6), 612–619.

Ferzli, R., & Karam, L. J. (2009). A no-reference objective image sharpness metric based on the notion of just noticeable blur (JNB). *IEEE Transactions on Image Processing, 18*(4), 717–728.

Fessler, J. A., & Booth, S. D. (1999). Conjugate-gradient preconditioning methods for shift-variant PET image reconstruction. *IEEE Transactions on Image Processing, 8*(5), 688–699.

Fleming, A. D., Philip, S., Goatman, K. A., Olson, J. A., & Sharp, P. F. (2006). Automated assessment of diabetic retinal image quality based on clarity and field definition. *Investigative Ophthalmology & Visual Science, 47*(3), 1120–1125.

Gao, H. (2012). Fast parallel algorithms for the x-ray transform and its adjoint. *Medical Physics, 39*(11), 7110–7120.

Ghrare, S. E., Ali, M. A. M., Ismail, M., & Jumari, K. (2008). Diagnostic quality of compressed medical images: Objective and subjective evaluation. In *International Conference on Modeling & Simulation, 2008, AICMS 08*. Second Asia.

Giancardo, L., Abramoff, M. D., Chaum, E., Karnowski, T. P., Meriaudeau, F., & Tobin, K. W. (2008). Elliptical local vessel density: A fast and robust quality metric for retinal images. In *30th Annual International Conference of the IEEE Engineering in Medicine and Biology Society, 2008*.

Ginesu, G., Massidda, F., & Giusto, D. D. (2006). A multi-factors approach for image quality assessment based on a human visual system model. *Signal Processing: Image Communication, 21*(4), 316–333.

Gonzalez, A. B. D., & Darby, S. (2004). Risk of cancer from diagnostic X-rays: Estimates for the UK and 14 other countries. *Lancet, 363*(9406), 345–351.

Goossens, B., Luong, H., Platiša, L., & Philips, W. (2012). Optimizing image quality using test signals: Trading off blur, noise and contrast. In *4th International Workshop on Quality of Multimedia Experience, Yarra Valley, VIC, Australia* (pp. 260–265). Piscataway, USA: IEEE.

Grills, I. S., Hugo, G., Kestin, L. L., Galerani, A. P., Chao, K. K., Wloch, J., et al. (2008). Image-guided radiotherapy via daily online cone-beam CT substantially reduces margin requirements for stereotactic lung radiotherapy. *International Journal of Radiation Oncology Biology Physics, 70*(4), 1045–1056.

Haacke, E. M., Mittal, S., Wu, Z., & Neelavalli, J. (2009). Susceptibility-weighted imaging: Technical aspects and clinical applications, part 1. *American Journal of Neuroradiology, 30*(1), 19–30.

Han, X., Pearson, E., Bian, J., Cho, S., Sidky, E. Y., Pelizzari, C. A., & Pan, X. (2010). Preliminary investigation of dose allocation in low-dose cone-beam CT. In *NSS/MIC: IEEE Nuclear Science Symposium & Medical Imaging Conference, Record* (pp. 2051–2054). Knoxville, TN.

Han, X., Pearson, E., Pelizzari, C., Al-Hallaq, H., Sidky, E. Y., Bian, J., et al. (2015). Algorithm-enabled exploration of image-quality potential of cone-beam CT in image-guided radiation therapy. *Physics in Medicine & Biology, 60*(12), 4601–4633.

Horie, N., Morikawa, M., Nozaki, A., Hayashi, K., Suyama, K., & Nagata, I. (2011). "Brush sign" on susceptibility-weighted MR imaging indicates the severity of moyamoya disease. *American Journal of Neuroradiology, 32*(9), 1697–1702.

Hoxworth, J., Lal, D., Fletcher, G., Patel, A., He, M., Paden, R., et al. (2014). Radiation dose reduction in paranasal sinus CT using model-based iterative reconstruction. *AJNR American Journal of Neuroradiology, 35*(4), 1–6.

Hua, Y., Liu, L., & Zhao, Q. (2015). Medical image quality assessment via contrast masking. In *8th International Congress on Image and Signal Processing (CISP), Shenyang, China* (pp. 964–968). Piscataway, USA: IEEE.

Iftekharuddin, K. M., Zheng, J., Islam, M. A., & Ogg, R. J. (2009). Fractal-based brain tumor detection in multimodal MRI. *Applied Mathematics and Computation, 207*(1), 23–41.

Islam, M. K., Purdie, T. G., Norrlinger, B. D., Alasti, H., Moseley, D. J., Sharpe, M. B., et al. (2006). Patient dose from kilovoltage cone beam computed tomography imaging in radiation therapy. *Medical Physics, 33*(6), 1573–1582.

Jaffray, D. A., Siewerdsen, J. H., Wong, J. W., & Martinez, A. A. (2002). Flat-panel cone-beam computed tomography for image-guided radiation therapy. *International Journal of Radiation Oncology Biology Physics, 53*(5), 1337–1349.

Jain, A. K., Duin, R. P. W., & Mao, J. (2000). Statistical pattern recognition: A review. *IEEE Transactions on Pattern Analysis and Machine Intelligence, 22*(1), 4–37.

Jensen-Kondering, U., & Böhm, R. (2013). Asymmetrically hypointense veins on T2* w imaging and susceptibility-weighted imaging in ischemic stroke. *World Journal of Radiology, 5*(4), 156–165.

Jin, K., Lu, H., Su, Z., Cheng, C., Ye, J., & Qian, D. (2017). Telemedicine screening of retinal diseases with a handheld portable non-mydriatic fundus camera. *BMC Ophthalmology, 17*(1), 89.

Karimi, D., Deman, P., Ward, R., & Ford, N. (2016). A sinogram denoising algorithm for low-dose computed tomography. *BMC Medical Imaging, 16*(1), 11.

Kawaguchi, A., Sharafeldin, N., Sundaram, A., Campbell, S., Tennant, M., Rudnisky, C., Weis, E., & Damji, K. F. (2017). Tele-ophthalmology for age-related macular degeneration and diabetic retinopathy screening: A systematic review and meta-analysis. *Telemedicine and E-Health*.

Keller, J. M., Gray, M. R., & Givens, J. A. (1985). A fuzzy k-nearest neighbor algorithm. *IEEE Transactions on Systems, Man, and Cybernetics, 15*(4), 580–585.

Khieovongphachanh, V., Hamamoto, K., & Kondo, S. (2008). Study on image quality for medical ultrasonic echo image compression by wavelet transform. In *International Symposium on Communications and Information Technologies (ISCIT 2008)* (pp. 160–165).

Kim, S., Yoshizumi, T. T., Frush, D. P., Toncheva, G., & Yin, F. F. (2010). Radiation dose from cone beam CT in a pediatric phantom: Risk estimation of cancer incidence. *AJR American Journal of Roentgenology, 194*(1), 186–190.

Kircher, M. F., de la Zerda, A., Jokerst, J. V., Zavaleta, C. L., Kempen, P. J., Mittra, E., et al. (2012). A brain tumor molecular imaging strategy using a new triple-modality MRI-photoacoustic-Raman nanoparticle. *Nature Medicine, 18*(5), 829–834.

Koopmans, P. J., Manniesing, R., Niessen, W. J., Viergever, M. A., & Barth, M. (2008). MR venography of the human brain using susceptibility weighted imaging at very high field strength. *Magnetic Resonance Materials in Physics, Biology and Medicine, 21*(1), 149–158.

Krupinski, E. A., & Jiang, Y. (2008). Anniversary paper: Evaluation of medical imaging systems. *Medical Physics, 35*(2), 645–659.

Lee, S. C., & Wang, Y. (1999). Automatic retinal image quality assessment and enhancement. *Proceedings of SPIE Image Processing, 3661,* 1581–1590.

Leng, S., Yu, L., Zhang, Y., Carter, R., Toledano, A. Y., & McCollough, C. H. (2013). Correlation between model observer and human observer performance in CT imaging when lesion location is uncertain. *Medical Physics, 40*(8), 081908.

Li, T., Li, X., Wang, J., Wen, J., Lu, H., Hsieh, J., et al. (2004). Nonlinear sinogram smoothing for low-dose X-ray CT. *IEEE Transactions on Nuclear Science, 51*(5), 2505–2513.

Li, Z., Yu, L., Trzasko, J. D., Lake, D. S., Blezek, D. J., Fletcher, J. G., et al. (2014). Adaptive nonlocal means filtering based on local noise level for CT denoising. *Medical Physics, 41*(1), 011908.

Lichy, M. P., Aschoff, P., Plathow, C., Stemmer, A., Horger, W., Mueller-Horvat, C., et al. (2007). Tumor detection by diffusion-weighted MRI and ADC-mapping—Initial clinical experiences in comparison to PET-CT. *Investigative Radiology, 42*(9), 605–613.

Liu, J., He, J., Chen, H., Ma, L., Zhang, Q., Pan, L. (2012). A comparative study of assessment methods for medical image quality. In *5th International Conference on Biomedical Engineering and Informatics (BMEI), Chongqing, China* (131–134). Piscataway, USA: IEEE.

Manduca, A., Yu, L., Trzasko, J. D., Khaylova, N., Kofler, J. M., McCollough, C. M., et al. (2009). Projection space denoising with bilateral filtering and CT noise modeling for dose reduction in CT. *Medical Physics, 36*(11), 4911–4919.

Mansouri, A., Aznaveh, A. M., Torkamani-Azar, F., & Jahanshahi, J. A. (2009). Image quality assessment using the singular value decomposition theorem. *Optical Review, 16*(2), 49–53.

Marrugoa, A. G., Millán, M. S., Šorel, M., Kotera, J., & Šroubek, F. (2015). Improving the blind restoration of retinal images by means of point-spread-function estimation assessment. In *Tenth International Symposium on Medical Information Processing and Analysis* (Vol. 9287, pp 92871D).

Matenine, D., Goussard, Y., & Després, P. (2015). GPU-accelerated regularized iterative reconstruction for few-view cone beam CT. *Medical Physics, 42*(4), 1505–1517.

McBain, C. A., Henry, A. M., Sykes, J., Amer, A., Marchant, T., Moore, C. M., et al. (2006). X-ray volumetric imaging in image-guided radiotherapy: the new standard in on-treatment imaging. *International Journal of Radiation Oncology Biology Physics, 64*(2), 625–634.

Morita, N., Harada, M., Uno, M., Matsubara, S., Matsuda, T., Nagahiro, S., et al. (2008). Ischemic findings of T2*-weighted 3-tesla MRI in acute stroke patients. *Cerebrovascular Diseases, 26*(4), 367–375.

Mucke, J., Möhlenbruch, M., Kickingereder, P., Kieslich, P. J., Bäumer, P., Gumbinger, C., et al. (2015). Asymmetry of deep medullary veins on susceptibility weighted MRI in patients with acute MCA stroke is associated with poor outcome. *PLoS ONE, 10*(4), e0120801.

Narvekar, N. D., & Karam, L. J. (2010). An improved no-reference sharpness metric based on the probability of blur detection. In *Workshop on Video Processing and Quality Metrics*.

Narvekar, N. D., & Karam, L. J. (2011). A no-reference image blur metric based on the cumulative probability of blur detection (CPBD). *IEEE Transactions on Image Processing, 20*(9), 2678–2683.

Neitzel, U., Gunther-Kohfahl, S., Borasi, G., & Samei, E. (2004). Determination of the detective quantum efficiency of a digital X-ray detector: Comparison of three evaluations using a common image data set. *Medical Physics, 31*(8), 2205–2211.

Ogawa, S., Lee, T. M., Kay, A. R., & Tank, D. W. (1990). Brain magnetic resonance imaging with contrast dependent on blood oxygenation. *Proceedings of the National Academy of Sciences, 87*(24), 9868–9872.

Othman, A. E., Brockmann, C., Yang, Z., Kim, C., Afat, S., Pjontek, R., et al. (2016). Impact of image denoising on image quality, quantitative parameters and sensitivity of ultra-low-dose volume perfusion CT imaging. *European Radiology, 26*(1), 167–174.

Pambrun, J., & Noumeir, R. (2013). Compressibility variations of JPEG2000 compressed computed tomography. In *35th Annual International Conference of the IEEE EMBS, Osaka, Japan* (pp. 3375–3378).

Paulus, J., Meier, J., Bock, R., Hornegger, J., & Michelson, G. (2010). Automated quality assessment of retinal fundus photos. *International Journal of Computer Assisted Radiology and Surgery, 5*(6), 557–564.

Ramirez-Giraldo, J. C., Trzasko, J., Leng, S., Yu, L., Manduca, A., & McCollough, C. H. (2011). Nonconvex prior image constrained compressed sensing (NCPICCS): Theory and simulations on perfusion CT. *Medical Physics, 38*(4), 2157–2167.

Reichenbach, J. R., Barth, M., Haacke, E. M., Klarhöfer, M., Kaiser, W. A., & Moser, E. (2000). High-resolution MR venography at 3.0 Tesla. *Journal of Computer Assisted Tomography, 24*(6), 949–957.

Samei, E., Ranger, N. T., Dobbins, J. T., III, & Chen, Y. (2006). Intercomparison of methods for image characterization. I. Modulation transfer function. *Medical Physics, 33*(5), 1454–1465.

Schuhbaeck, A., Achenbach, S., Layritz, C., Eisentopf, J., Hecker, F., Pflederer, T., et al. (2013). Image quality of ultra-low radiation exposure coronary CT angiography with an effective dose <0.1 mSv using high-pitch spiral acquisition and raw data-based iterative reconstruction. *European Radiology, 23*(3), 597–606.

Şevik, U., Köse, C., Berber, T., & Erdöl, H. (2014). Identification of suitable fundus images using automated quality assessment methods. *Journal of Biomedical Optics, 19*(4), 046006.

Sheikh, H. R., Sabir, M. F., & Bovik, A. C. (2006). A statistical evaluation of recent full reference image quality assessment algorithms. *IEEE Transactions on Image Processing, 15*(11), 3441–3452.

Shepp, L. A., & Logan, B. F. (1974). The Fourier reconstruction of a head section. *IEEE Transactions on Nuclear Science, 21*(3), 21–43.

Shnayderman, A., Gusev, A., & Eskicioglu, A. M. (2006). An SVD-based grayscale image quality measure for local and global assessment. *IEEE Transactions on Image Processing, 15*(2), 422–429.

Siddon, R. L. (1985). Fast calculation of the exact radiological path for a three-dimensional CT array. *Medical Physics, 12*(2), 252–255.

Sidky, E. Y., Duchin, Y., & Pan, X. (2011). A constrained, total-variation minimization algorithm for low-intensity X-ray CT. *Medical Physics, 38*(S1), S117–S125.

Sutha, V. J., & Latha, P. (2011). Wavelet based quality enhancement for medical images. In *International Conference on Recent Advancements in Electrical, Electronics and Control Engineering, Sivakasi, India* (pp. 277–280). Piscataway, USA: IEEE.

Szabo, T. L. (2004). Diagnostic ultrasound imaging: Inside out. Academic Press.

Szabo, T. L. (2004). Diagnostic ultrasound imaging: Inside out. Academic Press.

Tang, J., Nett, B.E., & Chen, G.H. (2009). Performance comparison between total variation (TV)-based compressed sensing and statistical iterative reconstruction algorithms. *Physics in Medicine & Biology, 54*(19): 5781.

Tian, P., Teng, I. C., May, L. D., Kurz, R., Lu, K., Scadeng, M., et al. (2010). Cortical depth-specific microvascular dilation underlies laminar differences in blood oxygenation level-dependent functional MRI signal. *Proceedings of the National Academy of Sciences, 107* (34), 15246–15251.

Toet, A., & Lucassen, M. P. (2003). A new universal colour image fidelity metric. *Displays, 24*(4), 197–207.

Tsai, D. Y., Lee, Y., & Matsuyama, E. (2008). Information entropy measure for evaluation of image quality. *Journal of Digital Imaging, 21*(3), 338–347.

Vaccaro, A. R., Madigan, L., Schweitzer, M. E., Flanders, A. E., Hilibrand, A. S., & Albert, T. J. (2001). Magnetic resonance imaging analysis of soft tissue disruption after flexion-distraction injuries of the subaxial cervical spine. *Spine, 26*(17), 1866–1872.

Wagner, R. F., Metz, C. E., & Campbell, G. (2007). Assessment of medical imaging system and computer aids: A tutorial review. *Academic Radiology, 14*(6), 723–748.

Wang, Z., Bovik, A. C., Sheikh, H. R., & Simoncelli, E. P. (2004). Image quality assessment: From error visibility to structural similarity. *IEEE Transactions on Image Processing, 13*(4), 600–612.

Wang, S., Ding, Y., Dai, H., Qian, D., Yu, X., & Zhang, M. (2014). Generalized relative quality assessment scheme for reconstructed medical images. *Bio-Medical Materials and Engineering, 24*(6), 2865–2873.

Wang, J., Li, T., Lu, H., & Liang, Z. (2006). Penalized weighted least-squares approach to sinogram noise reduction and image reconstruction for low-dose X-ray computed tomography. *IEEE Transactions on Medical Imaging, 25*(10), 1272–1283.

Wang, C., Song, R., Yerfan, J., Yang, L., Wang, S., Zhang, M., et al. (2016). A comparison study of single echo susceptibility weighted imaging and combined multi-echo susceptibility weighted imaging in visualizing asymmetric medullary veins in stroke patients. *PLoS ONE, 11* (8), e0159251.

Xu, Q., Yang, D., Tan, J., Sawatzky, A., & Anastasio, M. A. (2016). Accelerated fast iterative shrinkage thresholding algorithms for sparsity-regularized cone-beam CT image reconstruction. *Medical Physics, 43*(4), 1849–1872.

Xu, Q., Yu, H., Mou, X., Zhang, L., Hsieh, J., & Wang, G. (2012). Low-dose X-ray CT reconstruction via dictionary learning. *IEEE Transactions on Medical Imaging, 31*(9), 1682–1697.

Xue, W., Zhang, L., Mou, X., & Bovik, A. C. (2014). Gradient magnitude similarity deviation: A highly efficient perceptual image quality index. *IEEE Transactions on Image Processing, 23* (2), 684–695.

Yan, H., Cervino, L., Jia, X., & Jiang, S. B. (2012a). A comprehensive study on the relationship between the image quality and imaging dose in low dose CBCT. *Physics in Medicine & Biology, 57*(7), 2063–2080.

Yan, S., Sun, J. Z., Yan, Y. Q., Wang, H., & Lou, M. (2012b). Evaluation of brain iron content based on magnetic resonance imaging (MRI): comparison among phase value, R2* and magnitude signal intensity. *PLoS ONE, 7*(2), e31748.

Yan, H., Wang, X., Shi, F., Bai, T., Folkerts, M., Cervino, L., et al. (2014). Towards the clinical implementation of iterative low-dose cone-beam CT reconstruction in image-guided radiation therapy: Cone/ring artifact correction and multiple GPU implementation. *Medical Physics, 41* (11), 119912.

Yang, M. H., Kriegman, D. J., & Ahuja, N. (2002). Detecting faces in images: A survey. *IEEE Transactions on Pattern Analysis and Machine Intelligence, 24*(1), 34–58.

Yu, H., & Cai, Y. (2014). Contrast sensitivity function calibration based on image quality prediction. *Optical Engineering, 53*(11), 113107.

Zana, F., & Klein, J. C. (2001). Segmentation of vessel-like patterns using mathematical morphology and curvature evaluation. *IEEE Transactions on Image Processing, 10*(7), 1010–1019.

Zeileis, A., Smola, A., & Hornik, K. (2004). kernlab-an S4 package for kernel methods in R. *Journal of Statistical Software, 11*(9), 1–20.

Zhang, L., Cavaro-Ménard, C., Callet, P. L., & Ge, D. (2015). A multi-slice model observer for medical image quality assessment. In *International Conference on Acoustics, Speech and Signal Processing (ICASSP), South Brisbane, Australia* (pp. 1667–1671). Piscataway, USA: IEEE.

Zhang, L., Cavaro-Menard, C., Callet, P. L., & Tanguy, J. Y. (2012). A perceptually relevant channelized joint observer (PCJO) for the detection-localization of parametric signals. *IEEE Transactions on Medical Imaging, 31*(10), 1875–1888.

Zhang, Y., & Chandler, D. M. (2013). No-reference image quality assessment based on log-derivative statistics of natural scenes. *Journal of Electronic Imaging, 22*(4), 1–23.

Zhang, Y., Leng, S., Yu, L., Carter, R., & McCollough, C. H. (2014). Correlation between human and model observer performance for discrimination task in CT. *Physics in Medicine & Biology, 59*(13), 3389–3404.

Zhu, Y., & Ding, Y. (2017). Auto-optimized paralleled sinogram noise reduction method based on relative quality assessment for low-dose X-ray CT. *Journal of Medical Imaging and Health Informatics, 7*(1), 278–282.

Chapter 9
Challenge Issues and Future Work

Looking back on the last few decades, obstacles and challenges are appearing all the way along the progresses in the field of image quality assessment (IQA). In this chapter, we attempt to draw attention to some of the pressing challenges that are demanding solutions and are instructive for the research in the near future. Among these problems, which are come up with the best usage of our knowledge, some have been existed since the very beginning of this field, such as the pursuit of accuracy and efficiency, while the others are newly occurring due partly to the development of this and other related fields, and partly to the increasing demand of high-quality images in various applications. To some degree, it is the significant progress of IQA in recent years that allows us to find out more and more problems calling for solutions, which is a good sign of further development of IQA in both theory and practice. In the following, we would like to introduce these challenges and see how they can offer guides for our future work.

The issue that we want to introduce at the first is about subjective study. Using labeled data as the golden standard is a widely acknowledged experimental approach in scientific research. The appearance of subjective IQA databases in early 2000s has greatly facilitated objective study. However, there exists some undesirable uniqueness of the subjective experiments in IQA comparing to other fields. The main problem is that the subjective experiments are too consuming in time and labor and are not completely reliable because of the nonrepeatability. In other fields, take objective recognition as an example, through experiments, the golden standard can be almost 100% true. Yet for IQA, similar effects can only be achieved when highly complex experimental flows are organized. This fact leads to two unwanted results that the subjective findings across databases might vary and the scale of the databases is limited. Then, over-fitting problem easily occurs, which can be witnessed by the fact that the objective methods often perform drastically different across databases. Moreover, the limited scale makes machine learning hard to be implemented. Therefore, subjective IQA is likely to remain a very important topic in the future. Hopefully, databases with larger scale and more distortion types will be designed and provide more guidelines for objective study.

© Zhejiang University Press, Hangzhou and Springer-Verlag GmbH Germany 2018
Y. Ding, *Visual Quality Assessment for Natural and Medical Image*,
https://doi.org/10.1007/978-3-662-56497-4_9

Secondly, more quality-aware features are to be exploited, which is actually one of the most conventional topics in the history of IQA. The concept of quality-aware feature is abstract, to a certain degree. For human visual system (HVS), perceiving quality of images is through a complicated process based on the biological characteristics. However, by extracting features using mathematical tools, the simulation of this complicated process can be avoided. Thanks to previous works, extensive types of image features are demonstrated quality-aware, e.g., luminance, contrast, edges, textures, and corresponding tools effective in capturing these features are discovered. A modern opinion is that the most promising approach to make use of them is by ensemble, because the interests of HVS cannot be represented using single feature or single metric, with today's technology. In addition, many novel methods are applied to take advantage of the features in novel ways. For instance, constructing codebooks is a popular strategy appearing quietly recently for processing feature data. To conclude, the exploitation of unknown quality-aware features, as well as the re-organization of known ones, are one of the main directions of further objective IQA study.

Thirdly, the popularization of machine learning techniques has affected IQA research significantly. The most common stage where machine learning is adopted is pooling. With quality indices derived with the extracted quality-aware features (which is mostly several values rather than one) and the training target (the quality score), the pooling becomes a typical regression problem. The implementation of naive regression models such as linear regression and exponential regression can be found since the very early stage of IQA. With the development of machine learning techniques, especially in recent years with the wide acknowledgment of these techniques, complex models such as support vector machine (SVM) and neural network (NN) are also becoming common. The improving capability of the learning models arouses the concern that when an objective method is performing good, it is controversial that whether the feature extraction stage or the learning stage is to take credit. On the other hand, since the scale of training samples is quite small, it is not suitable to adopt complex regression models such as deep NN, to avoid over-fitting. Generally, extracting multiple quality-aware features and applying SVM for regression is a promising combination, because for training dataset at such scale, SVM is usually a better choice than NN. Also, another notice to avoid over-fitting is that the number of quality indices is also limited. A more serious concern about machine learning is that it has the potential to replace the role of feature extraction. Although currently the training data is far from sufficient for automatically "learn" how to extract features, convolutional NN (CNN) in object recognition has provided a good example of this possibility. Thus, the role machine learning can play for IQA is worth considering.

Fourthly, a trend of IQA is that its application scenario has already extended to many other fields other than the traditional planar (2D) natural images. For natural images, stereoscopic (3D) IQA is drawing a lot of attention in recent years, thanks to the development of 3D displaying techniques and the application in entertainment industry. Aside from natural images, IQA is already implemented for many application-specific areas. We have introduced medical IQA as one instance in this book, but the similar situation appears in many fields. The development of application-specific methods is a very good sign of IQA research, since its practical usages are confirmed. Application-specific IQA can be regarded as a generalization of natural IQA, thus many detailed processing schemes of the former can be derived from the latter. Thinking from a long term, IQA of each concrete field would possibly be enlightening for each other, together making IQA more advanced.

Fifthly, the specified IQA application scenarios lead to several issues that have never appeared before. Starting from square one, even the definition of IQA differs according to the specific situation. For natural images, the viewers of visual signal are common people, and the goal is to quantify the degree of visual distortions. Therefore, the objective methods are ought to correlate with characteristics of HVS and the ideal images contain certain natural statistical scenes (NSS). The logic is not alike for application-specific IQA. For instances, the viewers of medical images are doctors; the goal is to judge how good the images are for the purpose of diagnosis; the viewers of remote sensing images are geographers with the goal to decide how well they are for the analysis of topographic features, etc.; thus, neither HVS-based nor NSS-based natural IQA methods are likely to be straightforwardly applicable. Obviously, the difference in objects and targets influences the process in both designing subjective databases, where experimental setting can be drastically different and objective methods, where the captured feature types should be correspondingly different. Other differences should also be noted, for example, since application-specific methods aim at detailed scenarios, in many cases the reference images are unavailable, so full-reference and reduced-reference methods are not applicable. These differences are not to deny the commonalities between them, but they outline the necessity of studying these specific areas, and the exploitation of them is also an important topic in the future.

Sixthly, IQA-related biological and psychological research deserves attention. Findings from these areas have already provided lots of inspirations for IQA and other fields in computer vision. However, our current understanding upon HVS is still quite shallow. There is certain knowledge about HVS that we are safe to be sure of, including the properties of eyes, the properties of so-called simple and complex cells, the hierarchies in visual cortex, etc., yet the knowledge is far from adequate for us to build mathematical models to simulate how HVS functions. On the other hand, the straightforward simulation for HVS is too consuming in time and computational resources. There are already HVS-based objective IQA methods been proposed, of which the computational complexity is amazingly high. So, there is a compromise we have to make between the strict modeling of HVS and the pursuit of effectiveness. It is quite certain that the strict model cannot be constructed in the near future, but inspirations still can be drawn for IQA from novel findings in biology and psychology. Besides the low-level findings in biology and psychology, models in many other related areas can be introduced for IQA. For instance, the model of visual saliency is widely applied in objective IQA methods, which is actually an integration and abstraction of many properties of HVS. If we regard physical properties of HVS as the lowest abstraction level, and IQA is the desired application level, then the abstraction level of visual saliency is medium, between the highest and lowest level. The point is that the design of IQA methods is not necessarily directly based on the lowest or physical level; instead, it can be the employment of findings that are already highly abstracted. In 3D IQA, the models built to simulate binocular rivalry and visual discomfort have already shown effectiveness, for another two examples. Actually, similar instances can be easily found, such as the tools that we utilize to extract features.

Seventhly, the focus of IQA research is transferring to no-reference methods. In the very beginning of IQA study, most of objective methods are full-reference, but it is evident that no-reference methods are more robust for various application scenarios, and is therefore much more desirable. Later, there emerges a new type called reduced-reference, as a compromise between them. The difficulty of designing no-reference methods is evident, and the performance of them is not competitive to full-reference ones, in the early time. It is especially hard to construct no-reference IQA models capable of dealing with multiple distortion types, so there are also many distortion-specified no-reference methods proposed. But methods proposed in most recent years are mostly able to process images with arbitrary distortion types, and the ratio of no-reference methods is gradually increasing. Moreover, the predicting accuracy of many novel no-reference methods is superior than conventional full-reference methods. This trend denotes the overall improvement of the IQA field. In theory, the difficulty of designing no-reference methods has been greatly decreased, due to the findings of previous works. In practice, the progress in no-reference study makes IQA more feasible for real-world applications. So, the increasing trend of no-reference methods is unescapable.

Eighthly, to also make IQA methods more feasible for real-world applications, a recent trend is the study of color images. Many conventional methods deal with only the grayscale images; in other words, from the prospective of image channels, they merely take interests in the luminance channel. If it is color images whose quality we want to assess, there are basically two ways to decompose them. One way is to define three chromatic channels, usually red, green, and blue (RGB), and the luminance can be computed as the weighted average of each corresponding pixel. The other way is to use only one or two chromatic channels, in which case usually luminance takes up a single channel. Either way, the luminance only reflects part of information in an image. Although it has been found that gray scale aimed IQA methods may perform quite well for chrominance distortions, since the distortion for color reflects on the change of luminance, to certain degrees, the specific study for chrominance should be valued because in reality, grayscale images have become quite rare nowadays. It may worth well studying that how HVS perceives color.

Last but not least, computational efficiency is becoming a more and more important metric in the evaluation of objective IQA methods. The fact arouses more concern indicates that IQA is equipped for more real-world systems. In practice, accuracy and efficiency is almost always a trade-off, of which the latter is what lots of previous works exhibiting high correlation with HVS fail to achieve. Efficiency here can be interpreted in two aspects, time-wise and space-wise. The consuming time is usually the more attractive one, because it directly decides whether a method can be equipped in reality. Efficiency in space denotes how many computational resources it is required to implement objective an IQA method, which determines its feasibility for specific computational platforms. In computer science, it is usually encouraged to trade time with space. Thus, many parallel frameworks reporting higher time efficiency have been proposed. With the improvement of hardware performance, this thought is very promising. For practical usages, improving the efficiency of objective methods is a topic requiring lots of studying in the future.

IQA has evolved a lot in the last few decades. In academical research, standard experimental flows are set up, quality-aware features are discovered, properties of HVS are made use of, and the performance of objective methods is improved significantly. In application, IQA is generally accepted to be a necessary part in the framework of modern image and video processing systems and is equipped for applications in many generalized and specified areas. However, there are still rooms for improvement. To summarize, the gap between the current status of IQA and the real-world demanding still exists, especially with the latter develops unstoppably in pace with the advancing of electronic engineering and computer science. Therefore, how to make up this gap is the focus of IQA research in the foreseeable future.

Index

Printed in the United States
by Bookmasters

Printed in the United States
By Bookmasters